STUDIES IN NUMBER THEORY

Studies in Mathematics

The Mathematical Association of America

Volume 1: STUDIES IN MODERN ANALYSIS
edited by R. C. Buck

Volume 2: STUDIES IN MODERN ALGEBRA
edited by A. A. Albert

Volume 3: STUDIES IN REAL AND COMPLEX ANALYSIS
edited by I. I. Hirschman, Jr.

Volume 4: STUDIES IN GLOBAL GEOMETRY AND ANALYSIS
edited by S. S. Chern

Volume 5: STUDIES IN MODERN TOPOLOGY
edited by P. J. Hilton

Volume 6: STUDIES IN NUMBER THEORY
edited by W. J. LeVeque

W. J. LeVeque
University of Michigan

D. J. Lewis
University of Michigan

Julia Robinson
University of California, Berkeley

D. H. Lehmer
University of California, Berkeley

P. T. Bateman
University of Illinois

H. G. Diamond
University of Illinois

Studies in Mathematics

Volume 6

STUDIES IN NUMBER THEORY

W. J. LeVeque, editor
University of Michigan

Published by
The Mathematical Association of America

Distributed by
Prentice-Hall, Inc.

13–541359–1
Library of Congress Catalog Card Number 75–76868

Printed in the United States of America
Prentice-Hall, Inc., Englewood Cliffs, N.J.

Current printing (last digit):

10 9 8 7 6 5 4 3 2 1

CONTENTS

INTRODUCTION
W. J. LeVeque, 1

A BRIEF SURVEY OF DIOPHANTINE EQUATIONS
W. J. LeVeque, 4

DIOPHANTINE EQUATIONS: p-ADIC METHODS
D. J. Lewis, 25

DIOPHANTINE DECISION PROBLEMS
Julia Robinson, 76

COMPUTER TECHNOLOGY APPLIED
TO THE THEORY OF NUMBERS
D. H. Lehmer, 117

ASYMPTOTIC DISTRIBUTION OF BEURLING'S
GENERALIZED PRIME NUMBERS
P. T. Bateman and H. G. Diamond, 152

INDEX, *211*

STUDIES IN NUMBER THEORY

INTRODUCTION

W. J. LeVeque

The five articles in this volume by no means present a uniform coverage of the contemporary areas of research in number theory, but they do illustrate the remarkable breadth both of the subject itself and of the array of other mathematical theories that have successfully been brought to bear on arithmetical questions.

The first three articles are all on diophantine equations. The first, by myself, is an account without details of some of the main threads running through the still rudimentary theory of diophantine equations. It begins with a description of the classical results concerning rational and integral points on algebraic curves, up to about 1930; there was a hiatus of some 25 years thereafter when the subject was quiescent, and the more recent work is perhaps too intricate for inclusion in a volume of this sort. This is followed by a survey of results, mostly much more recent, concerning equations in many variables, and equations having an abnormally large number of solutions. I have been bold enough to put this article first, not because of its merit, but because it is the one requiring the least background, and because it serves to introduce the topics covered in much greater depth in the papers of D. J. Lewis and J. Robinson.

That by Lewis is concerned with the solvability of diophantine equations in p-adic fields, and with the application of p-adic methods to equations over the rational field. After a lucid exposition of the basic properties of the p-adic fields, the author gives a detailed analysis of the relationship between p-adic and rational solvability of linear and quadratic equations. This is followed by an account of the recent work on forms in many variables centering around the Artin conjecture. The remainder of the article is devoted to a complete and leisurely exposition of Skolem's method, by means of which rather precise information can sometimes be obtained on the number of solutions of specific diophantine equations.

Dr. Robinson's article is concerned with results obtained within the last decade, mostly by herself, H. Putnam, and M. Davis, on the decision problem for diophantine equations. After developing the necessary background on recursive functions, and the Davis normal form for primitive recursive functions, she shows the existence of exponential diophantine equations such that there is no method for deciding whether or not they are solvable. It is also shown that for any effectively given system of true axioms, there is a method for finding an exponential diophantine equation which has no solution but for which there is no proof from the axioms that the equation is unsolvable. Whether there are polynomial diophantine equations having similar properties is still unknown, but various elegant and tantalizingly simple-looking conditions are given in order that this should be the case.

In the fourth chapter, D. H. Lehmer combines a classification of a number of common arithmetical algorithms, according to their orders (lengths), with some pithy comments on the inadequacy of standard machine languages for number theoretic computations. The simplest algorithms, such as those for finding the G.C.D. of two integers, evaluating a Jacobi symbol or solving a quadratic congruence, are of order $\log n$, in the sense, for example, that the number of steps in the euclidean algorithm does not exceed a constant times the logarithm of the smaller of the two numbers. The determination of the continued fraction expansion of $\sqrt{n}$ is much more expensive, involving $O(\sqrt{n})$ steps, while wholesale listing, for example of the primes not exceeding n, may well be of order n itself. At the end of the paper, Lehmer points out how exact computations seemingly involving algebraic numbers can sometimes be effected by the adroit use of vectors.

The final chapter, by P. T. Bateman and H. G. Diamond, is concerned with the asymptotic distribution of the so-called "generalized primes," originally introduced by A. Beurling. With an arbitrary sequence P of real numbers p_j with $1 \leqq p_1 \leqq p_2 \leqq \cdots$, $p_j \to \infty$, there is associated the multiplicative semigroup N whose elements are the products of powers of the p_j, and P bears much the same relationship to N as the ordinary primes bear to the natural numbers. N is called the system of generalized integers associated with the generalized primes P. The counting function $I_P(x)$ of N, which gives the number of elements of N not exceeding x, clearly reduces to $[x]$ in the classical case, for $x > 0$, and the counting function $\pi_P(x)$ of P then reduces to the ordinary function $\pi(x)$. The principal topic of the paper of Bateman and Diamond is the relationship between the functions π_P and I_P, the goal being the theorem of Beurling that if

$$I_P(x) = Ax + O(x \log^{-\gamma} x),$$

where A is a positive constant and $\gamma > \frac{3}{2}$, then

$$\pi_P(x) = \frac{x}{\log x} + o\left(\frac{x}{\log x}\right).$$

(This "generalized prime number theorem" is best possible, in the sense that the conclusion is false for a suitable sequence P for which $\gamma = \frac{3}{2}$.)

The articles are all self-contained, in the sense that they can be read without extensive prior knowledge of number theory. On the other hand, they abound with further references to the literature, so that they can be used to gain entry to these branches of the subject.

A BRIEF SURVEY OF DIOPHANTINE EQUATIONS

W. J. LeVeque

This article is devoted to a description, mostly informal, of some of the broad principles which have influenced research in the theory of diophantine equations during recent years. A much more detailed exposition of one of the lines of thought touched on here will be found in the article by D. J. Lewis in this volume, and references are made elsewhere in the present article to expository accounts of other developments.

The name diophantine equation goes back to the Greek mathematician Diophantos of Alexandria (*c.* 250 A.D.), who made a number of contributions to the subject. It is a somewhat misleading term, since it seems to imply that a particular type of equation is under consideration, whereas in fact what is special is the nature of the required solutions. Any equation in one or more unknowns is called a diophantine equation if one asks for solutions in integers, or sometimes in rational numbers; polynomial equations, such as the Fermat equation $x^n + y^n = z^n$, are among the most natural to consider, but other types of equations have been investigated as well. Here we shall concentrate almost exclusively on polynomial equations.

1. PRE-NINETEENTH CENTURY WORK

Because of the important role of geometry in ancient mathematics, it is perhaps not surprising that among the first equations studied was that associated with the Pythagorean theorem, $x^2 + y^2 = z^2$. The solutions (3, 4, 5) and (5, 12, 13) of this equation are well known, and one might wonder whether there are additional solutions in integers. In fact, there are infinitely many such solutions, all obtainable by the rather simple formula

$$x = a^2 - b^2, \qquad y = 2ab, \qquad z = a^2 + b^2, \tag{1.1}$$

in which a and b may be given arbitrary integer values. [For example, to obtain the solution (3, 4, 5), put $a = 2$ and $b = 1$.] The Pythagorean equation exhibits an ideal behavior: Every solution is given by (1.1), and every set (x, y, z) obtainable from (1.1) is a solution. Such a complete solution would be desirable in all cases, of course, but it can rarely be achieved.

Although the Pythagorean theorem and the arithmetical investigation of the associated equation have traditionally been attributed to the Greek schools of Pythagoras and Euclid (550–300 B.C.), recent historical research (see Ore's book, listed in the bibliography) has shown that the Babylonians knew the entire story at least as early as 1600 B.C., and indeed that they used solutions of the Pythagorean equation to construct crude trigonometric tables. It was clearly because of its geometric significance that the equation received attention so early, for it was not until several centuries after Christ that the conceptually much simpler first degree equation $ax + by = c$ began to appear in Greek, Arabic, and Chinese word puzzles, although the theory required for its solution is to be found in Euclid's *Elements.* This equation does not appear in the extant writings of Diophantos, possibly because he considered it trivial. Most of his problems required finding squares or cubes with certain properties, and many, though not all, hinged on the solution of the Pythagorean equation. His work showed extreme ingenuity, but more importantly, it demonstrated clearly, and for the first time, that here was an entire new area of investigation.

Diophantos' work was apparently unknown in Europe during the Middle Ages, but was discovered and given two translations in the sixteenth and seventeenth centuries. Fortunately, a copy of the second translation, by C. Bachet, came into the hands of P. Fermat, who devoted his prodigious powers to comprehending and extending it; his work may be said to mark the beginning of the modern theory of numbers. Among his many results, he showed that the equation $x^4 + y^4 = z^4$ has no solution in nonzero integers, he gave a complete analysis of the problem of representing integers as sums of two squares, and he claimed to have proved that every positive integer is the sum of four squares, although he did not publish a proof. (One was finally given by J. Lagrange, in 1770.)

In the 250 years following Fermat, such men as L. Euler, J. Lagrange, K. Gauss, and E. Kummer, as well as lesser mathematicians, made further notable contributions to the subject. Their results were almost entirely limited, however, either to quadratic equations, such as the so-called Pell equation, $x^2 - dy^2 = c$, or to equations of extremely special form, such as the Fermat equation $x^n + y^n = z^n$. Thus, although Kummer (1857) showed that the latter equation has no solutions in nonzero integers when n is smaller than 100, and although the difficulties he encountered and partially overcame led to profound developments in algebra and the theory of algebraic numbers which in turn proved to be powerful tools in attacking other diophantine equations, it must still be said that almost until the beginning of the present century there was very little in the way of a general theory of diophantine equations, and that even today such a theory is restricted almost exclusively to equations in two variables.

2. EQUATIONS IN TWO VARIABLES

In the period from 1890 to 1910, three papers appeared that marked a new era in the theory of binary (two-variable) diophantine equations. In the first two, by D. Hilbert and A. Hurwitz (1890) and H. Poincaré (1901), sophisticated concepts and methods from algebraic geometry occurred for the first time, while in the third, A. Thue (1909) showed how to obtain information about the solvability of such equations by studying instead the solutions of certain inequalities. Let us examine these ideas in some detail.

A polynomial equation $f(x, y) = 0$ is represented geometrically by a certain curve, if x and y are allowed to assume real values, or by a so-called Riemann surface if x and y may take complex values. In studying these curves it is frequently useful to make a change of variables so as to obtain an equation that is in some sense simpler than the original. For example, the equation $y^2 = x^3 + 3x^2 + 3x + 1$ reduces to $u^2 = z^3$ under the translation $z = x + 1$, $u = y$. Not every transformation is suitable, however, if information about the new curve is to have implications for the old; what is required is that there be a one-to-one correspondence between the points of the two curves, under the transformation. The simplest transformations of this kind are those of the form

$$\begin{aligned} x &= \phi(z, u), \\ y &= \psi(z, u), \end{aligned}$$

where ϕ and ψ are rational functions of z and u, with the special property that when $f(x, y) = 0$, also z and u are rational functions of x and y,

$$\begin{aligned} z &= \Phi(x, y), \\ u &= \Psi(x, y). \end{aligned}$$

Such transformations, called *birational transformations,* play a central role in the study of algebraic curves. In general, not only the form of the equation but even its degree may be changed under a birational transformation, but still it is reasonable to consider all those curves birationally equivalent to a given one as different representations of a single algebraic relationship. While the degree may change, there is a more esoteric number, called the *genus* of the curve, which remains unchanged under birational transformations. Its exact definition need not concern us; it suffices to know that there is a number having this invariance property, and that it is a nonnegative integer. (It is, of course, the topological genus of the Riemann surface associated with the equation.)

The curves of genus 0 are especially simple, because of the following property: If $f(x, y) = 0$ is of genus 0, then there are rational functions $g(t)$, $h(t)$ of another variable t such that for every complex value of t, $f(g(t), h(t)) = 0$. In other words, the curve $f(x, y) = 0$ can be written in parametric form as

$$x = g(t), \qquad y = h(t). \tag{2.1}$$

For example, the circle $x^2 + y^2 = 1$ is of genus 0, and is parametrized by the equations

$$x = \frac{1 - t^2}{1 + t^2}, \qquad y = \frac{2t}{1 + t^2} \tag{2.2}$$

in the sense that for every choice of t, these equations determine a point on the circle, and every point on the circle can be obtained by choosing t suitably in (2.2). This example is somewhat special, however, in that not only the coefficients in the original equation but also those in the parametric equations are integers; for the general curve of genus 0 the functions $g(t)$ and $h(t)$ might have irrational coefficients, even when those in $f(x, y)$ are integers.

Hilbert and Hurwitz were the first to recognize the broad importance of these geometric ideas in arithmetical questions. Both of the sets of equations (1.1) and (2.2) had long been known, of course, and it was obvious that (1.1) results from (2.2) by choosing for t a rational value a/b and renaming the resulting rational numbers x and y as x/z and y/z with x, y, and z integers, but, possibly because this led to *rational* solutions of $x^2 + y^2 = 1$ rather than solutions in integers, no one before Hilbert and Hurwitz had systematically exploited the parametrizability of curves of genus 0.

They showed that every curve of genus 0, with rational coefficients, can be reduced by a birational transformation with rational coefficients to an equation of either first or second degree; it follows that all rational points, if such exist, on an arbitrary curve of genus 0 can be determined if the same can be done for first and second degree curves, and this is relatively easy. The question of solutions in integers remained open, since its solution required the later results of Thue.

Poincaré continued the investigation along these lines, and gave a more detailed analysis of the conditions under which a curve of genus 0 will have on it infinitely many, or only finitely many, rational points. He also considered curves of genus 1, and here too a significant reduction of the problem resulted; he showed that if such a curve has a rational point on it, then its equation can be reduced by a birational transformation with rational coefficients to the form

$$y^2 = 4x^3 + ax + b. \tag{2.3}$$

This fell far short of resolving the problem entirely, however, for even today the rational solutions of (2.3) are shrouded in mystery, although much has been learned in the meantime. The most fecund part of Poincaré's work, however, lay not in these rather specific results but in a much more tentative section, in which he indicated how the theory of algebraic functions and Abelian integrals might be brought to bear on diophantine problems, for curves of positive genus. This is a severely technical area, but, very crudely, the idea is as follows.

Although only curves of genus 0 have rational parametrizations, there is an analogous representation for curves of positive genus, in which one must (1) allow transcendental functions in place of rational functions, and (2) consider simultaneously a set of g points, if the curve is of genus g, rather than a single point (x, y) as in (2.1). The values of the parameters which yield a certain set of g points can be regarded as "coordinates" of the set, and it turns out that if two such sets are rational, in a suitable sense, then the set of points whose coordinates result by adding termwise the coordinates of the two sets is also rational; one calls the third set the sum of the other two. In the case of a curve of genus 1, a set of g points is a single point, and one can generate further rational solutions of the diophantine equation simply by adding together coordinates of known solutions and finding the points corresponding to the new coordinates. Poincaré commenced this kind of analysis of curves of genus 1, and prophesied that the same idea would be important for curves of higher genus. How right he was will appear as we proceed.

Thue's thinking was along entirely different lines. He concerned himself with diophantine equations of the form

$$ax^n + bx^{n-1}y + \cdots + hy^n = k, \quad a \neq 0, \tag{2.4}$$

in which the left-hand side is homogeneous of degree n and the coefficients $a, b, \cdots, h, k$ are integers, and began by factoring the polynomial on the left:

$$a(x - \alpha y)(x - \beta y) \cdots (x - \eta y) = k.$$

Here $\alpha, \beta, \cdots, \eta$ need not be rational numbers, of course, but as solutions of the algebraic equation $az^n + bz^{n-1} + \cdots + h = 0$ they are *algebraic* numbers. Suppose for simplicity that they are distinct from one another. The factored equation can be rewritten as

$$ay^n\left(\frac{x}{y} - \alpha\right)\left(\frac{x}{y} - \beta\right) \cdots \left(\frac{x}{y} - \eta\right) = k. \tag{2.5}$$

Now suppose that (2.4) has infinitely many solutions in integers x, y; then since, as is easily shown, there are at most n values of x corresponding to any fixed value of y, it follows that if the solutions are arranged in a sequence $(x_1, y_1), (x_2, y_2), \cdots$, necessarily the numerical value of y_m must become large as m grows indefinitely. Thus the factor ay^n on the left-hand side of (2.5) becomes large, and since the product of this factor with the others has the constant value k, it must be that some other factor becomes very small, that is, very close to 0. This means, in turn, that x/y must be a very good approximation to one of the numbers $\alpha, \beta, \cdots, \eta$. If it is α to which x/y is close, a little calculation shows that an inequality of the form

$$\left|\frac{x}{y} - \alpha\right| < \frac{A}{y^n} \tag{2.6}$$

must hold, where A is a certain constant depending on the coefficients $a, b, \cdots, h, k$. Thue therefore reasoned that one might be able to show that (2.4) cannot have infinitely many integral solutions by showing that for *every* algebraic number α which is a solution of an algebraic equation of degree n, and for every constant $A > 0$, the inequality (6) has only finitely many solutions in integers x, y. This inequality is different from the kind one meets in elementary mathematics in that the further one searches for solutions of it, the larger will be the denominator y, and the smaller will be the allowable error A/y^n. It is an example of a problem in *diophantine approximation*, and a difficult one at that. Thue gave a rather intricate proof that it has only finitely many solutions when $n > 2$, thereby establishing the same limitation on the solutions of (2.4) and at the same time showing the importance of this new way of attacking diophantine equations; his result, while not the first concerning equations of arbitrary degree and genus, was certainly the most general and elegant that had been obtained up to that time. Moreover, his result brought into clear focus one of the central difficulties of the subject: if an equation has only finitely many solutions, it is impossible to describe them by simple formulas like (2.1), for such formulas automatically furnish infinitely many solutions; rather, whatever solutions do exist are apt to be entirely fortuitous, and accidental events are difficult to analyze logically. It is possible, for example, that there is a single solution of the Fermat equation for some very large value of n; if so, its existence would naturally rule out any proof that the equation has no solutions, and would at the same time be incredibly difficult to predict on theoretical grounds.

The next major development in the arithmetical theory of curves came in 1922, when L. Mordell attempted to show that the equation

$$y^2 = ax^4 + bx^3 + cx^2 + dx + e,$$

which is of genus 1 except when $a, \cdots, e$ satisfy certain conditions (in which case it is of genus 0), has only finitely many solutions in integers. Some years later he wrote, "I thought at one time that I had proved the result, but there was a difficulty in the proof which held me up for some six months. However, on reconsidering my process and results, I noticed that I had really done something of an entirely different order of magnitude, which I probably could not have done if I had set out with that object." † Recall that Poincaré had shown how to "add" two points p_1 and p_2 on a curve of genus 1 to obtain a third point $p_3 = p_1 + p_2$; thus one can also obtain $p_3 + p_2 = p_1 + 2p_2$, and more generally one gets all the rational points $rp_1 + sp_2$, where r and s are integers. More generally still, one can start with m rational points $p_1, p_2, \cdots, p_m$ and obtain the further rational points $r_1p_1 + r_2p_2 + \cdots + r_mp_m$, where the r_i are integers, but it might happen that no matter how one chooses the points $p_1, p_2, \cdots, p_m$, there are rational points which cannot be obtained from them in this fashion. What Mordell had in fact shown was that if the curve has a cubic or quartic equation with integer coefficients, and is of genus 1, this cannot happen: there is invariably a finite collection of rational points on the curve such that every rational point can be expressed as a linear combination of them. Poincaré himself had conjectured this "finite basis" theorem, but he had no proof.

Seven years later A. Weil, in his doctoral dissertation, vastly extended Mordell's theorem by showing that it holds for arbitrary curves of genus 1 whose equations have algebraic coefficients, and indeed for points of a somewhat more general type than the rational points (namely, for points having coordinates in a fixed algebraic number field). Moreover, he obtained the analogous result for curves of higher genus, in which "rational point" is replaced by "rational set of g points." The proof itself, 35 pages in length, was by far the deepest application of algebraic geometry that had then been made to arithmetical questions, and was no doubt part of Weil's motivation in later reconsidering the foundations of algebraic geometry.

Very soon after the appearance of Weil's work on rational points, C. Siegel published the definitive paper on points with integer coordinates on curves, in which he effected a truly masterful fusion of the geometric and approximative techniques discussed above. In 1921 he had considerably strengthened Thue's results on the inequality (2.6), proving what came to be called the Thue-Siegel theorem, and had made a number of new applications to dio-

† *A Chapter in the Theory of Numbers.* Cambridge University Press, 1947.

phantine equations: now he combined the Thue-Siegel theorem with Weil's result on the one hand to show that no curve of positive genus has infinitely many points on it with integral coordinates, and with the Hilbert-Hurwitz-Poincaré analysis on the other hand to give an explicit characterization of those curves of genus 0 which have infinitely many such points. This completely settled the question of integral points on curves, except for detailed consideration of the number and nature of solutions of individual equations. Unfortunately, the proofs of both the finite basis theorem and Siegel's theorem are too technical to describe even crudely here.

Since 1930 there has been relatively little further progress in this area. The problem that towers above all else is to prove or disprove a conjecture of Mordell, that there are only finitely many *rational* points on a curve of genus larger than 1. It is an easy consequence of the technique of addition of points on a curve that there are curves of genus 1 having infinitely many rational points, but when $g > 1$ one must work with rational sets of g points instead, a rational *point* being the very special case in which the g points of a rational set are all identical with one another. Although there certainly are curves having infinitely many rational sets of points, it seems quite unlikely that infinitely many of these sets can collapse into single points, each taken g times. Thus Mordell concluded that there are probably only finitely many such points. The only significant progress on this problem was made quite recently by D. Mumford, who showed that the rational points are at any rate exceedingly rare when $g > 1$; in terms of a certain measure of the complicatedness of a rational point p, called the *height* $h(p)$, he showed that the number of rational solutions with $h(p) < x$ is at most $A \log x + B$, where A and B are constants depending only on the equation of the curve. In particular, his result implies that the number of solutions of the Fermat equation $x^n + y^n = z^n$ with $|z| < Z$ is at most $A \log \log Z + B$. An analogue of Mordell's conjecture has been proved in the so-called "function-field" case, but this sheds no light on the original question. Siegel's theorem too has been generalized in a variety of ways by K. Mahler, S. Lang and others, but again not in ways that are useful for Mordell's problem.

3. RATIONAL POINTS ON CURVES OF GENUS 1

Since curves of higher genus have proved so intractable, attention has been focused almost exclusively on curves of genus 1 during the last ten or fifteen years, and since there is rather little left to say about the points with integer coordinates, the emphasis has been on the rational points. In view of the Mordell-Weil theorem, all such points (if any exist) are expressible as combinations of a certain finite set of them (the "generators"), and one can now ask more sophisticated questions: how can one decide when a curve has rational points on it, and if so how many generators are there, and how can they be found? The proof of the Mordell-Weil theorem provides no answers to these questions, but simply shows the existence of the generators. The recent literature on these questions is large and exceedingly technical, and it will be possible here only to sketch the developments in broad outlines.

One powerful method of testing whether a given equation has rational solutions is to ask whether it has solutions in some larger field containing the field of rational numbers; if the answer is no, then there clearly cannot be rational solutions. This might seem to be an unpromising way to attack the problem because it would appear to be easier to find solutions in a large field than in a small one, but this is not always the case. For example, the equation $x^2 + 2y^4 = -1$ has no rational solutions because it has no real solutions, inasmuch as squares and fourth powers of real numbers are always nonnegative. Besides the real field, there are others, called the p-adic fields, which contain the rational field and in which it is sometimes quite easy to show that an equation has no solution. Without attempting an exact definition of a p-adic field (see the accompanying article by Lewis), we shall say only that an equation $f(x_1, x_2, \cdots, x_n) = 0$ has a p-adic solution, where p is a prime number, if and only if, for every positive integer N, no matter how large, there are integers $a_1(N), a_2(N), \cdots, a_n(N)$ such that $f(a_1(N), a_2(N), \cdots, a_n(N))$ is divisible by p^N. Obviously, it is impossible that $f(x_1, \cdots, x_n) = 0$ if $f(x_1, \cdots, x_n)$ is never divisible by p^N, since zero is divisible by every nonzero integer. Thus the existence of a p-adic solution, for every p, is necessary if the equation is to have a solution in integers, just as is the existence of real solutions. For certain types of equations this also works the other way around: If there are real soutions, and p-adic solutions for every p, then the equation has a solution in integers; when this is the case, one says that the Hasse principle applies. This is known to be true, for example, for all quadratic equations

$$\sum_{i,j=1}^{n} a_{ij}x_ix_j = 0$$

and also for all curves of genus 0, but while it is a useful general principle, it is not always valid. H. Reichardt gave the example of the curve $x^4 - 17 = 2y^2$, of genus 1, which has p-adic solutions for all p, and real solutions, but no solution in integers, and since then (1942) many other examples have been given and the phenomenon has been studied intensively. Even when the principle applies, it might seem to be very difficult to use in a positive way—that is, to show that solutions in integers exist—since it would appear to require that p-adic solvability be demonstrated for each of the infinitely many primes. In fact, however, it is a consequence of a theorem of Lang and Weil that p-adic solvability is automatic for all sufficiently large primes, and thus one need examine only finitely many primes. For each prime, solvability can be decided one way or another in a finite number of steps, so that in theory the Hasse principle is completely effective when it applies. And since its validity is not restricted to equations in two variables, it can also be used for example to investigate the solvability in rational numbers, rather than in integers, of equations of curves of genus 1 (or higher). For instance, every rational solution of $x^3 + y^3 = 1$, say $x = p/r$, $y = q/r$, would lead to an integer solution of $p^3 + q^3 = r^3$, and vice versa, so that each is solvable in the way specified if and only if the other is.

If a finite number of rational points on a curve are known, then the p-adic method is of no use in showing that there are no others, for then there are usually infinitely many p-adic and real solutions. In this case a device invented by Fermat, the so-called "method of infinite descent" can sometimes be applied, and is especially useful for curves of genus 1. The first step is to show that there is a birational transformation τ of the curve into itself such that every rational point p on the curve is of the form $\tau(q)$, where q is another rational point. Then one finds a way of associating with each rational point p a certain positive integer $H(p)$ which, like the height mentioned earlier, in some way measures the complicatedness of the point, and is such that if $p = \tau(q)$ is different from any of the rational points already known, then $H(q) < H(p)$. Then if there were additional solutions, there would be one for which $H(p)$ is smallest—but this is impossible, since $H(q)$ is even smaller.

While this method is applicable in principle to an equation in any number of variables, or to curves of arbitrary genus, its success depends on finding a suitable transformation of the curve into itself and, among the curves, this is easy only for curves of genus 1. In that case, there is, for instance, the transformation which sends each point q into the point $p = 2q = q + q$, where the addition is that discussed earlier in connection with Poincaré's work; more generally, one can map the curve into itself by transforming q into mq, where m is an arbitrary positive integer. Similar transformations are possible for curves of arbitrary genus g, of course, except that g-tuples of points occur in place of single points, and in fact the method of descent was central in both Weil's and Siegel's work. In the case $g = 1$, however, the method can be pushed much further, and it has been the basis of extensive work in recent years by J. Cassels, B. Birch and H. Swinnerton-Dyer, J. Tate, S. Lang, I. Shafarevich and others. Here homological methods also play an important role; these are hardly capable of explanation without elaborate preparation, and they lead to results not easily translated into nontechnical language. It must therefore suffice to say that through the work of these and other mathematicians, some understanding of the nature and number of generators of the rational points is beginning to emerge; at the same time, the latter are in many ways still very mysterious. (See the survey article by Cassels listed in the bibliography.)

4. EQUATIONS IN MANY VARIABLES

As soon as one leaves the relative comfort of equations in two (rational or integral) variables, a vast and nearly unknown terrain opens out ahead. Some of its details have been charted—there are certainly hundreds, and probably thousands, of research articles on the subject—but no broad comprehension has resulted. As might be expected, some success has been achieved with special equations in three and four variables, but very little with those in seventeen variables. There have, however, been a number of very interesting developments in recent years along somewhat different lines: instead of considering an individual equation in several variables, one considers a whole family of equations of similar form, but having different numbers of variables, and asks whether those having sufficiently many variables must always be solvable, and if so, how many variables are necessary. This approach is exemplified by the conjecture of E. Waring (1770) that for each fixed exponent k, the equation

$$x_1^k + x_2^k + \cdots + x_s^k = n, \qquad x_1 > 0, \cdots, x_s > 0 \tag{4.1}$$

is solvable, for every positive integer n, if s is sufficiently large. Long before Hilbert's proof of this general statement (1909), Lagrange had shown that every positive integer is the sum of four squares; thus if $g(k)$ is the smallest value of s for which Equation (4.1) is solvable for every positive n, he proved that $g(2) \leqq 4$. Since $7 = 2^2 + 1^2 + 1^2 + 1^2$ is the sum of four but no fewer squares, the exact value of $g(2)$ is 4. It had also been known for some time that $g(3) < 21$ and $g(4) < 41$, and by the time Hilbert's paper appeared it was known that $g(k)$ is finite for $k < 10$, although the upper bounds obtained, for the larger values of k, appeared to be very weak in comparison with the values suggested by tables.

An important variant of Waring's problem arises if one asks for the smallest value of s for which Equation (4.1) is solvable for every *sufficiently large* n, rather than for every positive n.

This value of s, denoted by $G(k)$, is the same as $g(k)$ for $k = 2$ but in general is smaller than $g(k)$; 23, for example, requires nine cubes, so that $g(3) \geqq 9$, but it is known that $G(3) < 7$, and probably $G(3) < 5$. In a certain sense $G(k)$ is more intrinsic to the problem than $g(k)$, since a few small integers n may require abnormally many kth powers, thereby increasing the size of $g(k)$.

There is a vast literature on both versions of Waring's problem. The solution for $g(k)$ is now nearly complete, namely, the exact value can be computed for each $k \geqq 5$, and for all but finitely many positive integers k, $g(k) = 2^k + A - 2$, where A is the largest integer not exceeding $(\frac{3}{2})^n$. This result, which had been conjectured for all k by Euler in the eighteenth century, comes by finding a comparatively small number n which is representable only by many kth powers of 1's and 2's; the proof that for almost all k no larger n would actually require more kth powers was nearly completed by L. Dickson and S. Pillai, but there was one detail which could not be supplied until K. Roth (1955) provided a definitive improvement of Thue's theorem concerning the inequality (2.6). [Roth showed that (2.6) has only finitely many solutions if the exponent n is replaced by any constant larger than 2; Mahler made the application to $g(k)$.]

Questions concerning $G(k)$ are more subtle, and the answers are less complete, because genuinely deeper properties of the integers are involved. Only the values $G(2) = 4$ and $G(4) = 16$ are known exactly. G. Hardy and J. Littlewood conjectured that $G(k) < 2k + 1$ when k is not a power of 2 and that $G(k) < 4k$ when it is, but the nearest result to this is Vinogradov's, that $G(k) < ck \log k$ for some constant c (for example, $c = 16$ is large enough). The technique, the details of which do not lend themselves to simple description, is to develop an approximate formula for the number of solutions of (4.1), and to deduce that the actual number of solutions is positive when s is sufficiently large. [In this simple form the idea works only for $s > 2^k + 1$, yielding $G(k) < 2^k + 1$, and to obtain the more precise result one must consider a special class of solutions of (4.1) which behaves rather more regularly.] There are two components in the principal term of the approximate formula, one factor corresponding to the "area" of the surface in n-dimensional space defined by Equation (4.1), and thus to the "quantity" of real solutions of (4.1), and the other factor corresponding to the number of p-adic solutions of (4.1), for the various different primes p. These factors emerge in the proof; p-adic solvability is not introduced a priori as was the case for equations in two variables. Since the approximate formula consists of the principal term plus an error term, it is of course necessary to verify that the factor corresponding to p-adic solutions, which does not vary with n, is different from zero; this is possible exactly when s is larger than the Hardy-Littlewood conjectural upper bound for $G(k)$. The still larger value of s occurring in the Vinogradov theorem is necessary in order to prove that the error term is of smaller order of magnitude than the principal term.

The special form (4.1) occurring in Waring's problem is subject to almost endless variation. One can allow negative values of $x_1, \cdots, x_s$ (which makes a difference only when k is odd, of course), or arbitrary signs between the terms, or integral coefficients in the terms, or each power x_i^k can be replaced by a polynomial of degree k in x_i. As long as the variables remain separated, in the sense that no two of them occur in the same term, and as long as all occur to the same highest degree, methods founded on similar ideas can be used to resolve these questions. As one example, we quote a recent theorem of H. Davenport and D. Lewis (1963): Let $c_1, \cdots, c_s$ be nonzero integers, not all of the same sign if k is even. Then if $s > k^2 + 1$, the equation $c_1x_1^k + c_2x_2^k + \cdots + c_sx_s^k = 0$ has infinitely many solutions in nonzero integers $x_1, \cdots, x_s$. In this case, in contrast to Waring's problem, it is the p-adic component of the question which provides the lower bound on s, rather than the associated formula for the number of real solutions. The lower bound is exact, in the sense that there are infinitely many values of k for which the equation in question does not have infinitely many solutions if $s < k^2 + 1$. (For a readable account of this technique, see Davenport's lecture notes, listed in the bibliography.)

The case in which there is a term in the equation involving at least two different variables is essentially more difficult. Prior to 1944, there were only two theorems of any generality dealing with this case. A. Meyer proved in 1884 that every quadratic form with integer coefficients assumes the value 0 for integer values of the variables $x_1, x_2, \cdots, x_n$, not all zero, if $n > 5$ and the form vanishes for *real* values of the variables, not all zero. (This latter condition must always be imposed on homogeneous equations of even degree, since for example a sum of squares obviously can be 0 only when all the variables are zero. For forms of odd degree, this difficulty does not arise.)

A set of values of the variables for which a certain polynomial vanishes is called a *zero* of the polynomial, and since clearly $f(0, 0, \cdots, 0) = 0$ if $f(x_1, x_2, \cdots, x_n)$ has no constant term, it is customary to refer to $(0, 0, \cdots, 0)$ as the trivial zero of a form. In this terminology, Meyer's theorem asserts that a quadratic form in at least five variables has a nontrivial zero in integers if it has a nontrivial zero in real numbers. The proof of Meyer's theorem is quite different from those for Waring-type problems; no approximate formula for the number of solutions is involved, but rather one shows, in essence, that Hasse's principle applies, and that there are always p-adic zeros if $n > 5$. In the thirties, Meyer's theorem was refined by Siegel, who obtained an approximate formula for the number of solutions, but he based his investigation on special properties of quadratic forms not shared by forms of higher degree, so that none of this early work provided any clue as to how to deal with forms of degree $d > 2$.

It was generally felt, however, that a form of fixed degree, or even a system consisting of a fixed number of forms of fixed degrees (not necessarily the same), should have a zero in integers if it has a real zero and if there are sufficiently many variables involved. As to the number of variables required, E. Artin pointed out in the 1930's that for every degree d, there are so-called "norm forms" in d^2 variables which have only the trivial zero; these are forms like $x^2 - 2y^2 = (x - \sqrt{2}y)(x + \sqrt{2}y)$ which split into linear factors having coefficients in some domain intermediate between the rational numbers and the complex numbers, and whose vanishing at a nontrivial point would entail a rational relationship among certain numbers not so related. (If $x_0^2 - 2y_0^2 = 0$, for example, with x_0 and y_0 nonzero integers, then $\sqrt{2} = \pm x_0/y_0$, which is false since $\sqrt{2}$ is irrational.) It follows that for every set of degrees $d_1, d_2, \cdots, d_r$, there are forms $f_1, f_2, \cdots, f_r$ having those degrees and not having a common nontrivial zero, if there are at most $d_1^2 + d_2^2 + \cdots + d_r^2$ variables. Artin conjectured that in fact this number of variables, plus 1, is sufficient to guarantee the existence of a nontrivial zero, in integers or in p-adic numbers, so long as the system has a nontrivial real zero. Some credence was lent to this conjecture by Meyer's theorem, which was exactly the case $r = 1$, $d = 2$ (so that $d^2 + 1 = 5$), and the (p-adic) conjecture was proved for a single cubic form by Lewis (1951), and for two quadratics by V. Demjanov (1956). In 1944 R. Brauer proved a qualitative form of the p-adic Artin conjecture, namely that a system of forms of fixed degrees has a p-adic zero provided that the number of variables is sufficiently large. In 1957 Lewis proved that a cubic form in sufficiently many variables has a zero in integers, and the next year Birch extended this to an arbitrary system of forms, all of odd degrees. At about the same time, Davenport obtained the first quantitative result for forms of degree larger than two, showing that a cubic form in at least 32 variables has a nontrivial zero in integers, and he later repeatedly lowered the estimate until it now stands at 16 variables. His proofs have a very curious logical twist; he obtains an approximate formula for the number of zeros in which none of the values of the variables $x_1, x_2, \cdots, x_n$ exceeds a certain bound N, and this formula shows that the number of such zeros is large when N is, but all this is on the assumption that the form has *no* nontrivial zero! It follows that the form must have a nontrivial zero, but the validity of the approximate formula has not been established. The proofs also involve the classical technique used for Waring-type problems, and the more refined results on the number of variables required depend steadily more heavily on geometric notions.

A considerable number of results in a similar vein have been proved within the past ten years by Davenport, Lewis, Birch, Swinnerton-Dyer, G. Watson and others. Very recently, J. Ax and S. Kochen proved a close approximation to the p-adic Artin conjecture by showing that a system of forms with p-adic coefficients has a p-adic zero if $n > d_1^2 + \cdots d_r^2$, for all sufficiently large primes p, but their proof does not yield a computable estimate for the lower bound for admissible p's. Motivated by their work, Terjanian has given an example of a form of degree 4 in $18 > 4^2$ variables having no p-adic zero for $p = 2$, so that the Artin conjecture as originally stated is false, and the Ax-Kochen result is in a certain sense best possible. On the other hand, Davenport and Lewis (1966) have shown that two Waring-type cubic forms in 18 variables always have a common nontrivial zero in integers, whereas the Artin number is $3^2 + 3^2 + 1 = 19$, so that for these special forms, the Artin conjecture is also sometimes too weak.

5. EQUATIONS WITH MANY SOLUTIONS

In the case of curves (polynomials in two variables), it follows from Siegel's fundamental theorem that unless the curve is birationally equivalent to a straight line, integral points must be distributed on it very sparsely, in the sense that the number of them with $\max(|x|, |y|) < N$ is smaller than $(\log N)^k$, for some constant k. This fact lends some support to the principle that if an equation in a rather small number of variables has an unusually large number of solutions in integers, it must be that the equation reduces to a triviality by a suitable change of variables. A special case of this principle (and of Siegel's theorem) which has been known for about 50 years is that if $f(x)$ is a polynomial with integral coefficients, whose value is the square of an integer for every integer x, then $f(x)$ is itself the square of some polynomial $g(x)$. In different language, if the diophantine equation $y^2 = f(x)$ is solvable for every integer x, then it has a parametrization $x = t$, $y = g(t)$. In principle, Siegel's theorem completely covers the case of equations in two variables, but since his result is very deep and its implications are not always immediately apparent, it is of some interest to deal in a more elementary way with certain special cases, as has been done for the equation $y^m = f(x)$ by LeVeque. Some progress has also been made with equations in three or more variables, as for example in a theorem of Davenport, Lewis, and Schinzel (and also proved independently by G.

Shimura) to the effect that if the equation $f(x) = y^2 + z^2$ is solvable for a suitably large set of values of x, then $f(x) = g^2(x) + h^2(x)$ for some polynomials $g(x)$ and $h(x)$. More generally, the same result holds if $x^2 + y^2$ is replaced by certain of the norm forms mentioned earlier, and these forms may involve arbitrarily many variables. It is still an open question, however, whether every value of a polynomial can be the sum of two cubes without the polynomial itself being the sum of the cubes of two polynomials. The principle under discussion certainly does not extend, without restriction, to arbitrary equations, since the equation $f(x) = y_1^2 + y_2^2 + y_3^2 + y_4^2$ is solvable for every x no matter what $f(x)$ may be, this being an immediate consequence of the fact that every integer is representable as a sum of four squares; no conjecture has yet been advanced giving broad conditions under which the principle holds.

In the special case of equations of the type $g(x_1, \cdots, x_m) = h(y_1, \cdots, y_n)$, however, in which g and h are forms of the same degree and satisfy certain technical conditions, S. Chowla has made the following conjecture: If such an equation is solvable in the y's for each choice of the x's, and vice versa, then there is a linear transformation with integral coefficients and determinant ± 1 sending g into h, provided that the number of positive values $< x$ assumed by one of the forms is of smaller order of magnitude than x itself. No proof has yet appeared.

BIBLIOGRAPHY

1. Ax, J., and Kochen, S., "Diophantine problems over local fields, I, II," *Amer. J. Math.*, 87 (1965), 605–630, 631–648.

2. Cassels, J. W. S., "Diophantine equations with special reference to elliptic curves," *J. London Math. Soc.*, 41 (1966), 193–291.

3. Chowla, S., "Problems in elementary number theory," *J. Reine Angew. Math.*, 222 (1966), 71–74.

4. Davenport, H., *Analytic methods for Diophantine equations and Diophantine inequalities*. Ann Arbor Publishers, Ann Arbor, Mich., 1963.

5. Davenport, H., and Lewis, D. J., "Homogeneous additive equations," *Proc. Roy. Soc. London Ser. A* 274 (1963), 443–460.

6. ———, and Lewis, D. J., "Cubic equations of additive type," *Philos. Trans. Roy. Soc. London, A* 261 (1966), 97–136.

7. ———, Lewis, D. J., and Schinzel, A., "Quadratic Diophantine equations with a parameter," *Acta Arith.*, 11 (1965/66), 353–358.

8. Hilbert, D., and Hurwitz, A., "Über die diophantische Gleichungen vom Geschlecht Null," *Acta Math.*, 14 (1890), 217–224.

9. LeVeque, W. J., "On the equation $y^m = f(x)$," *Acta Arith.*, 9 (1964), 209–219.

10. Mahler, K., "On the fractional parts of the powers of a rational number, II," *Mathematika*, 4 (1957), 122–124.

11. Mordell, L. J., "On the rational solutions of the indeterminate equations of the third and fourth degrees," *Proc. Cambr. Philos. Soc.*, 21 (1922), 179–192.

12. Mumford, D., "A remark on Mordell's conjecture," *Amer. J. Math.*, 87 (1965), 1007–1016.

13. Ore, O., *Number theory and its history.* McGraw-Hill, New York, 1948.

14. Poincaré, H., "Sur les propriétés arithmétiques des courbes algébriques," *J. Math. Pures Appl.*, (5) 7 (1901), 161–233.

15. Roth, K., "Rational approximations to algebraic numbers," *Mathematika*, 2 (1955), 1–20; corrigendum, 168.

16. Siegel, C. L., "Über einige Anwendungen diophantischer Approximationen," *Abh. Preuss. Akad. Wiss. Phys.—Mat. Kl.*, No. 1 (1929).

17. Terjanian, G., "Un contre-exemple à une conjecture d'Artin," *C. R. Acad. Sci. Paris, Sér. A–B* 262 (1966), A 612.

18. Thue, A., "Über annäherungswerte algebraischer Zahlen," *J. Reine Angew. Math.*, 135 (1909), 284–305.

19. Weil, A., "L'arithmétique sur les courbes algébriques," *Acta Math.*, 52 (1928), 281–315.

DIOPHANTINE EQUATIONS: p-ADIC METHODS

D. J. Lewis †

1. INTRODUCTION

In elementary number theory one encounters the problem of determining whether a specific equation, or inequality, has a solution in integers. The simplest example is that of a nonhomogeneous linear polynomial equation with integer coefficients. It is well known that such equations have an integral solution exactly when the greatest common divisor of the coefficients of the variables divides the constant term. Equally well known is the problem of determining all right triangles having integer sides, which is equivalent to determining all integer solutions, other than 0, 0, 0, of the equation $x^2 + y^2 = z^2$. In both of these examples, if an integer solution exists, there are infinitely many such solutions and they can be parametrized. On the other hand

† The author was partially supported by a grant from the National Science Foundation.

the equations $x^2 + 3y^2 = 1$, $x^2(x - 1)^2 = 2y(y - 1)$ and $2^y = x^2 + 7$ have only finitely many integral solutions; the integer solutions of the first have $y = 0$, those of the second have $y = 0, \pm 1, 2, -8, 9$, while those of the third have $y = 3, 4, 5, 7$, and 15. The inequalities $xy \leqq 1$, $x^2 + y^2 \leqq 1$ and $2x^2 + (5y - 3)^2 \leqq 1$ are examples where we have infinitely many, finitely many, and no integral solutions, respectively. The equation $x^3 - 117y^3 = 5$ is known to have at most 18 integral solutions but the exact number is unknown. There is a binary operation on the integral solutions of $x^3 = y^2z + 7z^3$ such that the solution set forms an abelian group. Of course, there is the obvious extension where we examine the existence of common integral solutions for a finite set of equations and inequalities. Problems in which the essential ingredient is the existence of integral solutions, are traditionally referred to as *diophantine problems*. The above examples illustrate typical qualitative and quantitative descriptions of solution sets for such problems.

Since the ring of integers **Z** is a subset of the field of real numbers **R**, diophantine problems may be viewed geometrically. In that case the equations and inequalities define surfaces and regions in real euclidean space and we seek to determine the existence and cardinality of the lattice or grid points (i.e., points with integral coordinates) on the intersection of the various surfaces and regions. Viewed in this way, one expects that most systems of equations and inequalities will have no integral points and that those which do should possess special properties. When our equations are homogeneous polynomials set equal to 0, it is customary and appropriate to visualize the surfaces as lying in projective space. When this is done, the tuple having all its coordinates 0 is not included in the set of solutions.

In this chapter we shall restrict our attention to diophantine equations. These equations can be put in the form of an expression (or function) set equal to zero, where the expressions are obtained from the integers and finitely many indeterminates by addition, subtraction, multiplication, and exponentiation. Given such an expression f, in m indeterminates, we shall say that an ordered

m-tuple is a *zero of* f provided it is a solution of $f = 0$. Similarly a zero of a set of expressions is a zero of each expression. In keeping with the convention mentioned above, a zero of a set of homogeneous polynomials must have at least one nonzero coordinate.

2. REALITY CONDITION

As we have already noted, the integral zeros of a set of expressions $\{f_\nu\}$ are the integral lattice points lying on the intersections of the various surfaces $f_\nu = 0$. In particular, the existence of a common integral solution of a set of equations implies that the intersection of the set of corresponding surfaces is nonempty; i.e., the set of equations has a real solution. Thus if one of the expressions is positive definite, or negative definite, then the set has no real and hence no integral zeros; e.g., $\sum_{j=1}^{m} x_j^2$ has no integral zeros. On the other hand, the polynomial $2x^2 + (5y - 3)^2 - 1$ obviously has real zeros, but no integral zeros. Summarizing, we have:

A. *A necessary, but not sufficient, condition that a set of equations have a common integral solution is that they have a common real solution.*

More generally, let $\mathfrak{R}$ be a commutative ring containing $\mathbf{Z}$ as a subring, and let $\mathfrak{R}^{(m)}$ denote the set of m-tuples with coordinates from $\mathfrak{R}$. If the expressions f_ν define functions from $\mathfrak{R}^{(m)}$ to $\mathfrak{R}$, as they would if the f_ν were polynomials over $\mathbf{Z}$, then the set of integral zeros of $\{f_\nu\}$ is a subset of the set of zeros of $\{f_\nu\}$ in $\mathfrak{R}^{(m)}$. In particular, if there are at most N zeros in $\mathfrak{R}^{(m)}$, then there are at most N integral zeros. Thus, considering appropriate rings $\mathfrak{R}$, we can get information concerning integral zeros.

Should $\mathfrak{R}$ be an algebraically closed ring, then every set of $r < m$ polynomials in m indeterminates has a zero in $\mathfrak{R}^{(m)}$. Finding integral zeros of polynomials is a difficult task exactly because $\mathbf{Z}$ is a very long way from being algebraically closed—in fact $\mathbf{Z}$

does not even contain the zeros of all linear polynomials over **Z**. The value of the reality condition is that while **R** is not algebraically closed, the ring $\mathbf{C} = \mathbf{R}[\sqrt{-1}\,]$ is. Hence, while the existence of real zeros is not a triviality, if they exist there are usually a great many of them and the reality condition is quickly verified.

3. CONGRUENCE CONDITIONS

B. *If a set of equations $\{f_\nu = 0\}$ defined over* **Z** *has an integral solution, then for every positive integer M the congruences $\{f_\nu \equiv 0 \pmod{M}\}$ have a solution. If the f_ν are homogeneous polynomials, then the greatest common divisor of the coordinates of at least one solution of the congruences is prime to M.*

Thus we have infinitely many necessary conditions for the existence of an integral solution, one for each positive integer M. These conditions are useful for demonstrating the nonexistence of integral solutions. As you might expect there is some artistry in choosing the appropriate modulus. For instance, $x_1^6 + x_2^6 + x_3^6 + x_4^6 = 5$ has no integral solution since it has no solution modulo 7.

A more sophisticated use of this idea shows that the equation $y^2 = x^3 + 7$ has no integral solution. If it did, then $y^2 \equiv x^3 + 7 \pmod{8}$ and x must be odd [otherwise $y^2 \equiv 7 \pmod{8}$], say $x = 2z + 1$. Then

$$\begin{aligned} y^2 + 1 &= x^3 + 8 = (x+2)(x^2 - 2x + 4) \\ &= (x+2)[\,(x-1)^2 + 3] = (2z+3)(4z^2+3). \end{aligned}$$

Now every integer of the form $4N + 3$ has a prime divisor q of that form, and hence $y^2 + 1 \equiv 0 \pmod{q}$ would be solvable for such a q dividing $4z^2 + 3$, which it is not.

If the f_ν are polynomials, we needn't consider all possible M, for we have

C. *If $M = p_1^{e_1} p_2^{e_2} \cdots p_s^{e_s}$, the polynomial congruences $\{f_\nu \equiv 0 \pmod{M}\}$ have a common solution if and only if for each i the congruences $\{f_\nu \equiv 0 \pmod{p_i^{e_i}}\}$ have a common solution.*

Proof: Suppose $a_{1i}, \cdots, a_{mi}$ is a common solution of the congruences $\{f_\nu \equiv 0 \pmod{p_i^{e_i}}\}$. Since the p_i are distinct primes, the Chinese remainder theorem implies the existence of $a_1, \cdots, a_m$ such that

$$a_{ti} \equiv a_t \pmod{p_i^{e_i}}, \quad \text{for } i = 1, \cdots, s \quad \text{and} \quad t = 1, \cdots, m;$$

hence for each i $\{f_\nu(a_1, \cdots, a_m) \equiv 0 \pmod{p_i^{e_i}}\}$, and therefore we have $\{f_\nu(a_1, \cdots, a_m) \equiv 0 \pmod{M}\}$. If for each i, the greatest common divisor of $a_{1i}, \cdots, a_{mi}$ is prime to p_i, then the greatest common divisor of $a_1, \cdots, a_m$ is prime to M. The proof in the other direction is obvious.

Our congruence conditions result in the study of zeros with coordinates in the finite quotient rings $\mathbf{Z}/p^n$. For any given prime p, we can find the zeros in $\mathbf{Z}/p^n$ by simply testing each of the finitely many m-tuples with coordinates in $\mathbf{Z}/p^n$. While in some cases this may be all that can be done, one naturally seeks a theory that avoids excessive computation. When $n > 1$, the rings $\mathbf{Z}/p^n$ have divisors of zero and this fact is a hindrance in developing a theory. On the other hand, if p is a prime, then $\mathbf{Z}/p$ is a finite field and C. Chevalley [10; see also 12, page 55] has shown that finite fields are nearly algebraically closed. Chevalley's theorem states:

Every homogeneous polynomial over a finite field F of degree d in $m > d$ variables has a zero in F.

Also, many questions concerning zeros in $\mathbf{Z}/p$ are resolvable by means of Gaussian sums.

Whenever possible one would prefer to disregard all extraneous congruence conditions. The following result, based on methods of Newton and Hensel, is helpful in this matter.

D. *Let $f_1, \cdots, f_r$ be polynomials with integer coefficients. If $a_1, \cdots, a_m$ are integers such that $\{f_\nu(a_1, \cdots, a_m) \equiv 0 \pmod{p}\}$ and the vectors $(u_{11}, \cdots, u_{1m}), \cdots (u_{r1}, \cdots, u_{rm})$, where $u_{ij} = (\partial f_i/\partial x_j)(a_1, \cdots, a_m)$, are linearly independent modulo p, then for each $T \geq 1$ the set of congruences $\{f_\nu \equiv 0 \pmod{p^T}\}$ has a solution $a_{1T}, \cdots, a_{mT}$, where $a_j \equiv a_{jT} \pmod{p}$, $j = 1, \cdots, m$.*

Proof: The proof is by induction on T. By hypothesis the result is true for $T = 1$. Suppose the result is true for $T \leqq N$; then $f_\nu(a_{1N}, \cdots, a_{mN}) = p^N b_{\nu N}$ and $\partial f_i/\partial x_j(a_{1N}, \cdots, a_{mN}) \equiv u_{ij} \pmod{p}$. Put $a_{j,N+1} = a_{jN} + p^N y_i$, $j = 1, \cdots, m$. Then

$$f_\nu(a_{1,N+1}, \cdots, a_{m,N+1}) \equiv f_\nu(a_{1N}, \cdots, a_{mN}) + p^N \sum_{j=1}^{m} u_{\nu j}y_j (\text{mod } p^{N+1})$$
$$\equiv p^N(b_{\nu N} + \sum_{j=1}^{m} u_{\nu j}y_j)(\text{mod } p^{N+1}).$$

Since the matrix (u_{ij}) has rank r (modulo p), the set of linear congruences $b_{\nu N} + \sum_{\nu=1}^{m} u_{\nu j}y_j \equiv 0(\text{mod } p)$ is soluble, and consequently we have the desired result for $T = N + 1$.

It should be noted that some condition on the Jacobian $(\partial f_\nu/\partial x_j)$ is needed in (D). For, in general, the solubility of $\{f_\nu \equiv 0(\text{mod } p)\}$ need not imply the solubility of $\{f_\nu \equiv 0(\text{mod } p^T)\}$ for every $T \geqq 1$. The congruence $x^2 + y^2 - 3 \equiv 0 \ (\text{mod } 3)$ is soluble, but $x^2 + y^2 - 3 \equiv 0 \ (\text{mod } 9)$ is not. Hence, we cannot do away with the congruence conditions for composite moduli entirely. For any given set of polynomials there are only finitely many congruence conditions for composite moduli which are not implied by congruence conditions with smaller moduli. However, it is usually difficult to specify those composite moduli that are essential.

There are various refinements of the above proposition. Typical of these refinements is the following: *If f is a polynomial with coefficients in $\mathbf{Z}$ and if $a_1, \cdots, a_m$ are integers such that $f(a_1, \cdots, a_m) \equiv 0(\text{mod } p^{2e+1})$, $\partial f/\partial x_j(a_1, \cdots, a_m) \equiv 0(\text{mod } p^e)$ for $j = 1, \cdots, m$, and $\partial f/\partial x_1(a_1, \cdots, a_m) \not\equiv 0(\text{mod } p^{e+1})$, then for each $k > e \geqq 1$ the congruence $f \equiv 0(\text{mod } p^k)$ is soluble.*

4. THE RING $\mathfrak{O}_p$

Our preceding discussion has indicated the desirability, the need, and in some sense the possibility of finding an equivalent set of conditions to replace the congruence conditions for composite moduli. Since each $\alpha \in \mathbf{Z}/p^n$ can be uniquely represented by an integer of the form

$$a_0 + a_1p + \cdots + a_{n-1}p^{n-1}, \quad 0 \leqq a_i \leqq p - 1,$$

there is a natural ring homomorphism θ_n of $\mathbf{Z}/p^n$ onto $\mathbf{Z}/p^{n-1}$ such that the representative of $\alpha\theta_n$ is $a_0 + \cdots + a_{n-2}p^{n-2}$. Let $\mathfrak{S}_p$ be the unrestricted cartesian sum $\mathbf{Z}/p + \mathbf{Z}/p^2 + \cdots + \mathbf{Z}/p^n + \cdots$

and let $\mathfrak{O}_p = \{\alpha = (\alpha_1, \alpha_2, \cdots) \in \mathfrak{G}_p \mid \alpha_n\theta_n = \alpha_{n-1}\}$. Clearly, $\mathfrak{O}_p$ is a subring of $\mathfrak{G}_p$. If α, β are nonzero elements of $\mathfrak{O}_p$ then there is an integer N such that $\alpha_N \neq 0 \neq \beta_N$. In that case, $\alpha_{2N}\beta_{2N} \neq 0$; whence $\alpha\beta \neq 0$. Thus $\mathfrak{O}_p$ has no divisors of zero. There is a homomorphism φ of $\mathbf{Z}$ into $\mathfrak{O}_p$, where the nth coordinate of $a\varphi$ is the residue class modulo p^n containing a. The map φ is injective, since $a\varphi = b\varphi$ implies that for every $n \geqq 1$, p^n divides $a - b$, and hence $a = b$. Thus $\mathfrak{O}_p$ is an integral domain of characteristic 0. Henceforth we shall view $\mathbf{Z}$ as being imbedded in $\mathfrak{O}_p$.

To each $n \geqq 1$ there is the homomorphism φ_n mapping $\mathfrak{O}_p$ onto $\mathbf{Z}/p^n$ where $\alpha\varphi_n = (\alpha_1, \alpha_2, \cdots)\varphi_n = \alpha_n$. Then $\varphi_n\theta_n = \varphi_{n-1}$. It follows that to each $\alpha \in \mathfrak{O}_p$ we can associate a unique infinite formal series $\sum_{\nu=0}^{\infty} a_\nu p^\nu$, where $\sum_{\nu=0}^{n-1} a_\nu p^\nu$ is the representation of $\alpha\varphi_n = \alpha_n$. Conversely, each such series determines an element of $\mathfrak{O}_p$. Thus there is a one-to-one map of $\mathfrak{O}_p$ onto the set of formal power series $\sum_{\nu=0}^{\infty} a_\nu p^\nu$. We should caution the reader that if $\sum a_\nu p^\nu + \sum b_\nu p^\nu = \sum c_\nu p^\nu$, one does not usually have $c_\nu = a_\nu + b_\nu$. We define the *order* of a nonzero element α in $\mathfrak{O}_p$ to be

$$\text{ord } \alpha = \text{ord } p^\alpha = \min \{n \mid a_n \neq 0\} = \min \{n \mid \alpha\varphi_n \neq 0\}.$$

Define ord $0 = \infty$. We assume, as usual, that if $n \in \mathbf{Z}$ then $\infty > n$, $\infty + n = \infty$, and $\infty + \infty = \infty$. It is easily verified that

1. $\text{ord } \alpha\beta = \text{ord } \alpha + \text{ord } \beta$,

2. $\text{ord } (\alpha + \beta) \geqq \min \{\text{ord } \alpha, \text{ord } \beta\}$.

Suppose $\alpha, \beta \in \mathfrak{O}_p$ and $\alpha\beta = 1$, then ord α + ord $\beta = 0$. Since $\alpha \neq 0 \neq \beta$, ord α and ord β are nonnegative integers and hence both are 0. Conversely, suppose $\alpha \in \mathfrak{O}_p$ and ord $\alpha = 0$. Then $\alpha\varphi_n$ is prime to p, for $n = 1, 2, \cdots$, and hence there exist unique integers β_n, v_n such that $(\alpha\varphi_n)\beta_n + p^n v_n = 1, 0 \leqq \beta_n < p^n$. Clearly, $\beta_n\theta_n = \beta_{n-1}$. The element $\beta = (\beta_1, \beta_2, \cdots)$ belongs to $\mathfrak{O}_p$, and $\alpha\beta = 1$. Thus the units of the ring $\mathfrak{O}_p$ are exactly those elements of $\mathfrak{O}_p$ of order 0.

Let $\mathfrak{p} = p\mathfrak{O}_p$ be the principal ideal in $\mathfrak{O}_p$ generated by the integer p, so that $\mathfrak{p}$ is exactly the set of nonunits of $\mathfrak{O}_p$. It follows that $\mathfrak{p}$ is a maximal ideal of $\mathfrak{O}_p$. If $\alpha, \beta \in \mathfrak{O}_p$ and $\alpha\beta \in \mathfrak{p}$, then $(\alpha\varphi_1)(\beta\varphi_1) \equiv 0(\text{mod } p)$, and hence at least one of α and β lies in $\mathfrak{p}$. Hence $\mathfrak{p}$ is a prime ideal of $\mathfrak{O}_p$. Suppose $\mathfrak{m}$ is a nonzero ideal of $\mathfrak{O}_p$. Let $I = \{\text{ord } \alpha \mid 0 \neq \alpha \in \mathfrak{m}\}$. I is a nonempty subset of positive integers and so has a least element, say k. Let γ be an element of $\mathfrak{m}$ such that ord $\gamma = k$, say $\gamma = b_k p^k + b_{k+1}p^{k+1} + \cdots = p^k(b_k + b_{k+1}p + \cdots)$, where $b_k \not\equiv 0(\text{mod } p)$. Since $b_k + b_{k+1}\, p + \cdots$ is a unit of $\mathfrak{O}_p$, it follows that $p^k \in \mathfrak{m}$ and $p^k\mathfrak{O}_p \subset \mathfrak{m}$. As each element of $\mathfrak{m}$ has order $\geqq k$, each element of $\mathfrak{m}$ lies in $p^k\mathfrak{O}_p$. Hence $\mathfrak{m} = p^k\mathfrak{O}_p$. Furthermore, we have $\mathfrak{m} = \mathfrak{p}^k$. Thus the only nonzero ideals in $\mathfrak{O}_p$ are the powers of the maximal prime ideal $\mathfrak{p}$. It follows that $\mathfrak{p}$ is the unique maximal ideal in $\mathfrak{O}_p$. It can be verified that there is a natural isomorphism of $\mathfrak{O}_p/\mathfrak{p}^k$ onto $\mathbf{Z}/p^k$.

5. ZEROS IN $\mathfrak{O}_p^{(m)}$

Let $\{f_\nu\}$ be a set of polynomials with coefficients in $\mathbf{Z}$. As noted in section 2, a necessary condition that $\{f_\nu\}$ has an integral zero is that it has a zero with coordinates in $\mathfrak{O}_p$. Such a zero implies that for each $k \geqq 1$, the set of congruences $\{f_\nu \equiv 0(\text{mod } p^k)\}$ is soluble. Furthermore, if some coordinate of the zero is a unit of $\mathfrak{O}_p^{(m)}$, then there are solutions of the congruences with a coordinate prime to p.

If for each $n \geq 1$ there is a solution $\alpha_{1n}, \cdots, \alpha_{mn}$ of $\{f_\nu \equiv 0\ (\text{mod } p^n)\}$ such that $\alpha_{jn}\theta_n = \alpha_{j,n-1}, j = 1, \cdots, m$ and $n = 1, 2, \cdots$, then $\alpha_1, \cdots, \alpha_m$, where $\alpha_j = (\alpha_{j1}, \alpha_{j2}, \cdots)$, is a zero of $\{f_\nu\}$ with coordinates in $\mathfrak{O}_p$. Consequently, if the set $\{f_\nu\}$ has no zeros in $\mathfrak{O}_p^{(m)}$ then to each solution $\alpha_{11}, \cdots, \alpha_{m1}$ of $\{f_\nu \equiv 0(\text{mod } p)\}$ there is an integer $n = n(\alpha_{11}, \cdots, \alpha_{m1})$ such that there is no solution of $\{f_\nu \equiv 0(\text{mod } p^n)\}$ whose image under $\theta_n\theta_{n-1}\cdots\theta_2$ is $\alpha_{11}, \cdots, \alpha_{m1}$. Since the set of solutions of $\{f_\nu \equiv 0(\text{mod } p)\}$ is finite, the set $\{n\}$ is bounded, say by N. It follows that $\{f_\nu \equiv 0(\text{mod } p^k)\}$ has no solution for $k > N$. For if $\alpha_{1k}, \cdots, \alpha_{mk}$ were such a solution, we would have $n(\alpha_{1k}\theta_k \cdots \theta_2, \cdots, \alpha_{mk}\theta_k \cdots \theta_2) \geq k > N$, a contradiction. If follows that if $\{f_\nu \equiv 0(\text{mod } p^k)\}$ is soluble for an infinite set of k's, then such is the case for all k and $\{f_\nu\}$ has a zero in $\mathfrak{O}_p^{(m)}$.

Furthermore, if for each k there is a zero with a coordinate prime to p, then there is a zero in $\mathfrak{O}_p^{(m)}$ having one coordinate a unit. We have therefore shown that the congruence conditions with a power of the prime p as modulus are equivalent to the existence of a zero with coordinates in $\mathfrak{O}_p$. Summarizing, we have

E. *A necessary condition that a set of polynomials $\{f_\nu\}$ over $\mathbf{Z}$ have a zero in $\mathbf{Z}^{(m)}$ is that they have a zero in $\mathbf{R}^{(m)}$ and, for each prime p, they have a zero in $\mathfrak{O}_p^{(m)}$.*

Working with the integral domain $\mathfrak{O}_p$ is advantageous, not only because the quantity of conditions is thereby reduced, but because $\mathfrak{O}_p$ has a very simple topology. This topology can be used effectively in studying the existence of zeros in $\mathfrak{O}_p^{(m)}$.

The process by which we constructed $\mathfrak{O}_p$ is very common in mathematics today. The ring $\mathfrak{O}_p$ is called the *inverse limit* of the rings $\mathbf{Z}/p^n$.

6. THE p-ADIC FIELD

Let W be the set of ordered pairs $\alpha \mid \beta$, where $0 \neq \beta$, $\alpha \in \mathfrak{O}_p$. The relation $\alpha \mid \beta \sim \gamma \mid \delta$ if and only if $\alpha\delta = \beta\gamma$ is an equivalence relation on W. Let $\mathbf{Q}_p$ be the set of equivalence classes of W together with the operations

$$\{\alpha \mid \beta\} + \{\gamma \mid \delta\} = \{\alpha\delta + \beta\gamma \mid \beta\delta\}, \; \{\alpha \mid \beta\}\{\gamma \mid \delta\} = \{\alpha\gamma \mid \beta\delta\}.$$

Then $\mathbf{Q}_p$ is a field, the field of quotients of the domain $\mathfrak{O}_p$. Since $\mathfrak{O}_p$ contains $\mathbf{Z}$, $\mathbf{Q}_p$ contains the field of rational numbers $\mathbf{Q}$. Each equivalence class contains an element of the form $\alpha \mid p^m$, where m is some nonnegative integer. If α is not a unit of $\mathfrak{O}_p$ and m is positive we can reduce the size of m. Thus each equivalence class of W contains a unique pair either of the form $p^k u \mid 1$ or $u \mid p^k$, where u is a unit of $\mathfrak{O}_p$ and $k \geq 0$. Since $u = \sum_{\nu=0}^{\infty} a_\nu p^\nu$, $0 \leqq a_\nu \leqq p - 1$, $a_0 \neq 0$, we see that $p^k u \mid 1$ can be represented by $p^k \sum_{\nu=0}^{\infty} a_\nu p^\nu$, while $u \mid p^k$ can be represented uniquely by $p^{-k} \sum_{\nu=0}^{\infty} a_\nu p^\nu$. Hence each nonzero element α of $\mathbf{Q}_p$ can be uniquely represented by a formal series $p^m \sum_{\nu=0}^{\infty} a_\nu p^\nu$, where $m \in Z$, $0 \leqq a_\nu \leqq p - 1$, $a_0 \neq 0$. As before, we call m the order of α.

Our series representations for elements of $\mathbf{Q}_p$ are merely formal sums. For aesthetic as well as utilitarian purposes it is desirable to introduce a metric on $\mathbf{Q}_p$ so that these series actually converge relative to the metric.

A *valuation* is a map, $a \rightarrow || a ||$, of a field F into the nonnegative real numbers having the following properties: For all $\alpha, \beta \in F$, and for some positive real c,

1. $||\alpha|| \geqq 0$, with equality holding if and only if $\alpha = 0$.
2. $||\alpha\beta|| = ||\alpha|| \cdot ||\beta|| \cdot$
3. If $||\alpha|| \leqq 1$, then $||1 + \alpha|| \leqq c$.

Clearly the absolute value function is a valuation on $\mathbf{Q}$, on $\mathbf{R}$, and on $\mathbf{C}$, with $c = 2$. The map $||\alpha||_p = p^{-\text{ord}\,\alpha}$ is a valuation on $\mathbf{Q}_p$ (and hence on $\mathbf{Q}$) with $c = 1$. Moreover, for this valuation on $\mathbf{Q}_p$,

4. $||\alpha + \beta||_p \leqq \max\{||\alpha||_p, ||\beta||_p\}$.

A valuation satisfying the more restrictive condition (4) is called a *nonarchimedean valuation.* The ordinary absolute value does not satisfy (4). If $c \leq 2$, then $|| \; ||$ satisfies the triangle inequality: $||\alpha + \beta|| \leq ||\alpha|| + ||\beta||$.

The map $|| \; ||_p$ is called the *p-adic valuation,* the field $\mathbf{Q}_p$ is called the *p-adic field* and the integral domain $\mathfrak{O}_p$ is called the *ring of p-adic integers.*

Let F be a field with a valuation $|| \; ||$. A sequence $\{a_n\}$ from F is said to be a *null sequence* (relative to $|| \; ||$) if $\lim_{n \to \infty} ||a_n|| = 0$. An element $b \in F$ is said to be the *limit* (relative to $|| \; ||$) of a sequence $\{b_n\}$ from F provided $\{b - b_n\}$ is a null sequence. A sequence $\{c_n\}$ from F is said to be a *cauchy sequence* (relative to $|| \; ||$) provided that for each $\epsilon > 0$ there is an integer $N(\epsilon)$ such that $||c_n - c_m|| < \epsilon$ whenever $m,n \geqq N(\epsilon)$. As usual, every sequence having a limit is a cauchy sequence.

Each element $p^m \sum_{\nu=0}^{\infty} a_\nu p^\nu$ of $\mathbf{Q}_p$ is a limit (relative to $|| \; ||_p$) of the sequence of rational numbers $s_n = p^m \sum_{\nu=0}^{n} a_\nu p^\nu$. Clearly not all such sequences have a limit in $\mathbf{Q}$. If $\{\gamma_n = p^{m_n} \sum_{\nu=0}^{\infty} c_{n\nu}p^\nu\}$ is a cauchy sequence of $\mathbf{Q}_p$, then there is an integer m such that $m_n = m$ for all sufficiently large n. Let $N_n = N(1/p^n)$ and put $\alpha_n = c_{N_n0} + c_{N_n1}p + \cdots + c_{N_nn}p^n$. Then $\alpha = (\alpha_1, \alpha_2, \cdots)$ is in $\mathbf{Q}_p$ and $p^m\alpha$ is the limit of the sequence $\{\gamma_n\}$. Thus we have

F. *Each cauchy sequence from $\mathbf{Q}_p$ has a p-adic limit in $\mathbf{Q}_p$ and each element of $\mathbf{Q}_p$ is a p-adic limit of a sequence of rational numbers.*

Apropos of this result, it may be illuminating to recall that relative to the absolute value, each cauchy sequence from $\mathbf{R}$ has a limit in $\mathbf{R}$ and each element of $\mathbf{R}$ is a limit of a sequence of rational numbers.

A valuation on a field F defines a Hausdorff topology, where the open neighborhoods of an element $a \in F$ are the sets

$$U_{a,\epsilon} = \{x \in F \text{ such that } ||x - a|| < \epsilon\}.$$

The field operations of addition and multiplication are continuous relative to this topology. Hence a polynomial function on F to F is a continuous function. It is easy to verify that the neighborhoods of $a \in \mathbf{Q}_p$ are exactly the cosets $a + \mathfrak{p}^m$, $m = 0, 1, \cdots$. These neighborhoods are both open and closed sets in the topological field $\mathbf{Q}_p$. By means of either Dirichlet's box principle or Tychonoff's theorem, one can show that any closed bounded subset A of $\mathbf{Q}_p$ is compact; i.e., given any covering of A by open sets, there is a finite subcollection of these open sets whose union covers A. Thus the sets in the fundamental system of neighborhoods of $\mathbf{Q}_p$ are both open and compact. In particular $\mathfrak{O}_p$ is open and compact. The domain $\mathfrak{O}_p$ is especially valuable since it is a compact set containing the ring of rational integers $\mathbf{Z}$; in fact, $\mathbf{Z}$ is dense in $\mathfrak{O}_p$. Closed, bounded subsets of $\mathbf{R}$ are also compact, but no compact subset of $\mathbf{R}$ contains $\mathbf{Z}$. Also, in contrast to the real numbers, it can be easily shown that $\sum_{n=1}^{\infty} \gamma_n$, $\gamma_n \in \mathbf{Q}_p$, converges in $\mathbf{Q}_p$ if and only if $\{\gamma_n\}$ is a null sequence. If a series from $\mathbf{Q}_p$ does converge, then any rearrangement of the series converges to the same limit as did the original series.

Suppose f is a continuous map of $\mathfrak{O}_p^{(m)}$ into $\mathfrak{O}_p$ and suppose that for each positive integer s there is a point $(\gamma_{1s}, \cdots, \gamma_{ms}) \in \mathfrak{O}_p^{(m)}$ such that $f(\gamma_{1s}, \cdots, \gamma_{ms}) \equiv 0 \pmod{\mathfrak{p}^s}$. Since $\mathfrak{O}_p$ is compact, so is $\mathfrak{O}_p^{(m)}$, and hence the set of points $\{(\gamma_{1s}, \cdots, \gamma_{ms})\}$ has at least one accumulation point in $\mathfrak{O}_p^{(m)}$. As a consequence, we see that the hypothesis of Proposition (D) implies the existence of a zero of $\{f_\nu\}$ in $\mathfrak{O}_p^{(m)}$. The argument, in the paragraph preceding (E), that the congruence conditions imply the existence of p-adic zeros was so involved because we did not then know that $\mathfrak{O}_p$ was compact. This result is a trivial consequence of the compactness of $\mathfrak{O}_p$.

Two valuations on a field F are said to be *equivalent* if they define equivalent topologies; i.e., each open neighborhood of one topology is contained in an open neighborhood of the other topology. It is relatively easy to show: *Distinct primes give rise to inequivalent valuations on* $\mathbf{Q}$ *and no p-adic valuation is equivalent to the absolute value.* It is not too difficult to show: *Each nontrivial valuation on* $\mathbf{Q}$ *is equivalent either to the absolute value or to one of the p-adic valuations.* Actually, inequivalent valuations on $\mathbf{Q}$ are highly independent, for one can prove:

G. Approximation Theorem: *Let* $\|\ \|_{p_1}, \cdots \|\ \|_{p_r}$ *be a finite set of inequivalent p-adic valuations on* $\mathbf{Q}$. *Suppose* $a_0, a_1, \cdots, a_r \in \mathbf{Q}$ *and* $e_0, e_1, \cdots, e_r$ *are positive real numbers, then there exists an* $a \in \mathbf{Q}$ *such that* $|a_0 - a| < e_0$, $\|a_n - a\|_{p_n} < e_n$, *for* $n = 1, \cdots, r$ *and* $\|a\|_p \leq 1$ *for all* $p \neq p_1, \cdots, p_r$.

Without the condition $|a_0 - a| < e_0$, this theorem is merely a reformulation of the Chinese remainder theorem. With this assumption a bit more care is needed in the proof.

We note four other results, of which the first two are simple exercises. The last two involve more effort.

1. If a is a rational number $\neq 0$, then $|a| \prod_{\text{all } p} \|a\|_p = 1$.

2. $\bigcap_{\text{all } p} [\mathfrak{O}_p \cap \mathbf{Q}] = \mathbf{Z}$.

3. Two valuations on a field are equivalent if and only if one is a positive real power of the other.

4. Each valuation on a field is equivalent to one satisfying the triangle inequality.

It follows from our discussion that $\mathbf{R}$ and the various $\mathbf{Q}_p$ are the only fields containing $\mathbf{Q}$ which are cauchy complete and contain no proper cauchy complete subfield. Such fields are called *local fields of* $\mathbf{Q}$. To emphasize this similarity, one often denotes $\mathbf{R}$ by $\mathbf{Q}_\infty$. In that case, one also sets $\mathfrak{O}_\infty = \mathbf{R}$.

We have passed lightly over some of the properties of local fields. For a complete development of this theory the reader should consult Artin [2], Mahler [27], O'Meara [30].

7. EQUIVALENT SETS OF POLYNOMIALS

Let $\mathfrak{O}$ be an integral domain and let $f_1, \cdots, f_r$ be polynomials in m indeterminates with coefficients in $\mathfrak{O}$. Let T be an invertible m by m matrix with coordinates in $\mathfrak{O}$ such that T^{-1} has coordinates in $\mathfrak{O}$. If $(x_1, \cdots, x_m) = (y_1, \cdots, y_m)T + (a_1, \cdots, a_m)$, $a_\nu \in \mathfrak{O}$, then $f_\nu(x_1, \cdots, x_m) = g_\nu(y_1, \cdots, y_m)$, where the g_ν have coefficients in $\mathfrak{O}$. Furthermore, if S is an r by r matrix with coordinates in $\mathfrak{O}$ with non-zero determinant such that $(g_1, \cdots, g_r)S = (b_1h_1, \cdots, b_rh_r)$, $0 \neq a \in \mathfrak{O}$, then either each of the sets $\{f_\nu\}$, $\{g_\nu\}$, and $\{h_\nu\}$ has a zero in $\mathfrak{O}$ or none of them do. The correspondence

$$\{f_1, \cdots, f_r\} \leftrightarrow \{h_1, \cdots, h_r\}$$

can be shown to be an equivalence relation, and sets of polynomials related in this fashion are said to be *equivalent* over $\mathfrak{O}$. If two sets of polynomials with coefficients in $\mathfrak{O}$ are equivalent over $\mathfrak{O}$, then one set has a zero in $\mathfrak{O}$ if and only if the other set has a zero in $\mathfrak{O}$. Hence in studying the existence of zeros, we can always pass to a more convenient equivalent set.

If the polynomials are forms, we modify the definition of equivalence by requiring each $a_\nu = 0$. Then the transformations take sets of forms into sets of forms and we have an equivalence relation on sets of forms. Two sets of forms with coefficients in $\mathfrak{O}$

are equivalent over $\mathfrak{O}$ if and only if they are equivalent over the field of quotients of $\mathfrak{O}$.

A set of polynomials over $\mathfrak{O}$ is said to be *degenerate over* $\mathfrak{O}$ if it is equivalent to a set with fewer variables. A set of polynomials $\{f_\nu\}$ over $\mathfrak{O}$ (of degree less than characteristic of $\mathfrak{O}$, when $\neq 0$) is degenerate over $\mathfrak{O}$ if and only if there exist $c_\mu \in \mathfrak{O}$, not all zero, such that $\sum_{\mu=1}^{m} c_\mu(\partial f_\nu/\partial x_\mu) = 0$, $\nu = 1, \cdots, r$. It follows from the theory of linear equations that a set of polynomials over $\mathfrak{O}$ (of degree less than characteristic of $\mathfrak{O}$, when $\neq 0$) is degenerate over some larger domain if and only if it is degenerate over $\mathfrak{O}$. A set of polynomials over $\mathfrak{O}$ which is degenerate over $\mathfrak{O}$ always has a zero in $\mathfrak{O}$. Hence in studying the zeros of sets of polynomials we can restrict our attention to nondegenerate sets.

8. LINEAR POLYNOMIALS

Any time we are given a set of necessary conditions for the existence of some object it is natural to ask if these conditions might also be sufficient. Since our necessary conditions for integral zeros entailed all the local fields of $\mathbf{Q}$, one has added reason to expect them to be sufficient. We now prove,

H. *A set of linear polynomials with coefficients in* $\mathbf{Z}$ *has a zero in* $\mathbf{Z}^{(m)}$ *if and only if it has a zero in each* $\mathfrak{O}_p^{(m)}$.

Proof: Because of (E) we need only prove that the local zeros imply an integral zero. A set of linear polynomials over $\mathbf{Z}$ is equivalent over $\mathbf{Z}$ to a set $\{f_\nu\}$, where

$$(8.1) \qquad \begin{cases} f_\nu = a_{\nu\nu}x_\nu + \sum_{\mu=\nu+1}^{m} a_{\nu\mu}x_\mu + b_\nu, & 1 \leqq \nu \leqq s, \\ f_\nu = b_\nu, & s < \nu \leqq r, \end{cases}$$

and $a_{11}a_{22}\cdots a_{ss} \neq 0$. If the original set of polynomials has a zero in $\mathfrak{O}_p$, so does this set. The existence of zeros in $\mathfrak{O}_p$ imply that $b_{s+1} = \cdots = b_r = 0$.

Suppose p does not divide $a_{11}\cdots a_{ss}$. Then for any $x_{s+1}, \cdots,$

$x_m \in \mathfrak{O}_p$ there exists unique $x_1, \cdots, x_s \in \mathfrak{O}_p$ such that $x_1, \cdots, x_m$ is a zero of (8.1).

Suppose $p|a_{11}\cdots a_{ss}$, say $\text{ord}_p(a_{11}\cdots a_{ss}) = k(p)$. Let $y_1^{(p)}, \cdots, y_m^{(p)}$ be a zero of (8.1) in $\mathfrak{O}_p^{(m)}$. Then for any $m - s$-tuple $x_{s+1}, \cdots, x_m$, such that $x_\nu \in y_\nu^{(p)} + \mathfrak{p}^{k(p)} \subset \mathfrak{O}_p$, there exists unique $x_1, \cdots, x_s \in \mathfrak{O}_p$ such that $x_1, \cdots, x_m$ is a zero of (8.1). Now by the approximation theorem (G) there exist $x_\nu \in \mathbf{Q}$, $\nu = s + 1, \cdots, m$, so that

$$||x_\nu - y_\nu^{(p)}||_p \leqq ||p^{k(p)}||_p, \quad \text{if } p|a_{11}\cdots a_{ss},$$
$$||x_\nu||_p \leq 1, \quad \text{if } p \nmid a_{11}\cdots a_{ss}.$$

These x_ν imply the existence of $x_1, \cdots, x_s \in \mathbf{Q}$ such that $x_1, \cdots, x_m$ is a rational zero of (8.1). But $x_1, \cdots, x_s$ were so chosen that $||x_\nu||_p \leqq 1$ for all primes and all ν. Hence the x_ν lie in $\mathbf{Z}$. Since (8.1) has an integral zero, so does the original set of linear polynomials.

9. QUADRATIC POLYNOMIALS

If $\alpha \in \mathbf{Q}_p$ is the square of an element of $\mathbf{Q}_p$ then ord α is even. Hence if $a \in \mathbf{Z}$ is a square in each p-adic field, then $a = \pm b^2$, for some $b \in \mathbf{Z}$. If a is also square in $\mathbf{R}$ then $a = b^2$. Thus we have,

I. *If $a \in \mathbf{Z}$, then $x^2 - a$ has a zero in $\mathbf{Z}$ if and only if it has a zero in each of the local rings $\mathfrak{O}_p$.*†

On the other hand, neither $(2x + 3)(5x + 7)$ nor $x^2 - 2xy + x - 3y - 1$ have integral zeros, although they do have zeros in each $\mathfrak{O}_p$. Thus, in general, the local integral conditions alone do not imply the existence of rational integral zeros for nonlinear polynomials. Although the two examples above do not have integral zeros they do have rational zeros. Hence we might ask if the existence of zeros in each of the $\mathbf{Q}_p$ always implies the existence of a rational zero. Such is indeed the case for a single quadratic polynomial.

† It can be shown that $x^2 - a$ has a zero in $\mathbf{Z}$ if it has a zero in all but finitely many $\mathfrak{O}_p$.

Let $f = Q + L + c$ be a nondegenerate quadratic polynomial in m indeterminates over $\mathbf{Q}$, where Q is a quadratic form, L is a linear form, and c is a constant. By "completing the square" one can show that f is equivalent over $\mathbf{Q}$ to either

$$\sum_{\nu=1}^{m} a_\nu x_\nu^2 + c \quad \text{or} \quad \sum_{\nu=1}^{m-1} a_\nu x_\nu^2 + x_m.$$

In the latter case, f obviously has a zero in $\mathbf{Q}$ and in each local field.

Suppose $f = \sum_{v=1}^{m} a_\nu x_\nu^2 + c$ is nondegenerate. Then $a_1 a_2 \cdots a_m \neq 0$. Put

$$F(x_0, x_1, \cdots, x_m) = cx_0^2 + a_1x_1^2 + \cdots + a_m x_m^2.$$

We shall show: f has a rational zero if and only if F does. If $c = 0$ then $f = F$; hence we suppose $c \neq 0$. If $(b_1, \cdots, b_m)$ is a zero of f then $(1, b_1, \cdots, b_m)$ is a zero of F, while if $(b_0, \cdots, b_m)$, with $b_0 \neq 0$, is a zero of F then $(b_1/b_0, \cdots, b_m/b_0)$ is a zero of f. Suppose $(0, b_1, \cdots, b_m)$ is a zero of F. Since F is homogeneous, some $b_i \neq 0$, say $b_m \neq 0$. Then $\sum_{\nu=1}^{m} a_\nu b_\nu^2 = 0$, $a_m b_m \neq 0$ and

$$\begin{aligned} F(y_0, b_1y_1, b_2y_1 + y_2, \cdots, b_my_1 + y_m) &= \\ &= cy_0^2 + y_1\Big(2 \sum_{\nu=2}^{m} b_\nu a_\nu y_\nu\Big) + \sum_{\nu=2}^{m} a_\nu y_\nu^2 \\ &= cy_0^2 + y_1 M + P. \end{aligned}$$

Now M is not the zero polynomial and hence there exist $c_2, \cdots, c_m$ such that $M(c_2, \cdots, c_m) \neq 0$. We can then choose c_1 so that $F(1, b_1c_1, \cdots, b_mc_1 + c_m) = 0$.

Thus when studying the rational zeros of a quadratic polynomial it suffices to study the rational zeros of a quadratic form.

10. ONE QUADRATIC FORM

For a single quadratic form we have the very satisfactory result:

J. *A quadratic form with coefficients in* $\mathbf{Q}$ *has a zero in* $\mathbf{Q}$ *if and only if it has a zero in each local field of* $\mathbf{Q}$.

Proof: A nondegenerate form over $\mathbf{Q}$ is equivalent over $\mathbf{Q}$ to a form $F = \sum_{\nu=1}^{m} a_\nu x_\nu^2$, where $a_1 = 1$, $a_1 \cdots a_m \neq 0$, the a_ν are square free rational integers and the greatest common divisor of any $m - 1$ of the a_ν is 1. Since $\mathbf{Q} \subset \mathbf{Q}_p$, a rational zero is always a zero in each $\mathbf{Q}_p$.

For the remainder of the proof, we suppose F has a zero in each local field and show F has a rational zero. The case $m = 2$ is Proposition (I).

The case $m = 3$: The proof is by contradiction. Among all forms of the above specified type which have zeros in all local fields but have no rational zero, choose one with $|a_2a_3|$ minimal. Suppose F is that form. On rearranging subscripts, we may assume $|a_2| \leqq |a_3|$. If $|a_3| = 1$, then $|a_2| = 1$ and since F has a zero in $\mathbf{R} = \mathbf{Q}_\infty$, at least one of a_2 and a_3 is -1, whence F has an obvious rational zero. Hence we must have $|a_3| \geq 2$.

Since F is homogeneous, there exist zeros in $\mathfrak{O}_p$ with at least one coordinate a unit. By applying the approximation theorem (G), we find there exist relatively prime integers x_0, y_0, z_0 such that

$$x_0^2 + a_2y_0^2 + a_3z_0^2 \equiv 0 \pmod{a_3^2}.$$

Let $d = (a_3, y_0)$. Then $d|x_0$; whence $d^2|a_3z_0^2$ and since a_3 is square free so is d and hence $d|z_0$. But $(x_0, y_0, z_0) = 1$, hence $d = 1$. Since $|a_3| \geqq 2$, it follows that $y_0 \neq 0$. Consequently there exists $w \in \mathbf{Z}$ such that $wy_0 \equiv x_0 \pmod{a_3}$, $|w| \leqq |a_3|/2$. Then

$$w^2 + a_2 = \frac{y_0^2w^2 + y_0^2a_2}{y_0^2} \equiv \frac{x_0^2 + y_0^2a_2}{y_0^2} \equiv 0 \pmod{a_3}.$$

Set $w^2 + a_2 = a_3b_3$. If $b_3 = 0$, then $(w, 1, 0)$ is a rational zero of F contrary to our assumption. Hence,

$$0 < |b_3| \leqq \frac{1}{|a_3|}\{w^2 + |a_2|\} \leq \frac{1}{|a_3|}\left\{\frac{1}{4}|a_3|^2 + |a_3|\right\}$$
$$\leq 1 + \frac{1}{4}|a_3|.$$

Since $|a_3| \geqq 2$, we have $0 < |b_3| < |a_3|$. Next observe that $(wx_1 + a_2x_2)^2 + a_2(x_1 - wx_2)^2 + b_3(a_3x_3)^2 = a_3b_3(x_1^2 + a_2x_2^2 + a_3x_3^2)$.

Hence F and $F^* = y_1^2 + a_2y_2^2 + b_3y_3^2$ are equivalent over $\mathbf{Q}$. Consequently F^* has zeros in each of the local fields but none in $\mathbf{Q}$. But $1 \leqq |a_2b_3| < |a_2a_3|$, contrary to the minimality assumption on $|a_2a_3|$. This proves the proposition for the case $m = 3$.

The case $m \geqq 4$: This case is more involved and in order to handle it we need to obtain some additional results for ternary forms. Consider the Hilbert map $(a, b)_p$ of ordered pairs of p-adic numbers, or real numbers, into $1, -1$ defined by

$$(a, b)_p = \begin{cases} 1, & \text{if the equation } ax^2 + by^2 = 1 \text{ has a solution in } \mathbf{Q}_p, \\ -1, & \text{otherwise.} \end{cases}$$

(Here, p may be ∞.) Clearly $(a, 1)_p = (a, -a)_p = (a, 1-a)_p = 1$, $(au^2, by^2)_p = (a, b)_p$ and $(a, a)_p = (a, -1)_p$, for each a and p.

Given a nonzero element $a \in \mathbf{Q}_p$ let $\mathcal{S} = \{0 \neq b \in \mathbf{Q}_p | (a, b)_p = 1\}$; then $\mathcal{S}$ contains all nonzero squares of $\mathbf{Q}_p$ and $b^{-1} \in \mathcal{S}$ if $b \in \mathcal{S}$. Suppose $ax^2 + by^2 = 1$ and $au^2 + cv^2 = 1$; then

$$\begin{aligned} bc(yv)^2 = (1 - ax^2)(1 - au^2) &= 1 - a(u^2 + x^2) + a^2(ux)^2 \\ &= (1 + aux)^2 - a(u + x)^2. \end{aligned}$$

If $1 + aux \neq 0$, then $b, c \in \mathcal{S}$ implies $bc \in \mathcal{S}$. The same conclusion holds if $1 + aux = 0$. Hence $\mathcal{S}$ is a multiplicative subgroup of $\mathbf{Q}_p^\times$, the nonzero elements of $\mathbf{Q}_p$, and $\mathcal{S}$ contains the group of nonzero squares $\mathbf{Q}_p^2$.

When $p = \infty$, $\mathcal{S}$ is the positive real number when $a < 0$ and is R^x when $a > 0$ and hence $[\mathbf{Q}_p^\times : \mathcal{S}] \leqq 2$. Now suppose that p is a finite prime. If $\alpha \in \mathbf{Q}_p^2$, then ord α is even and

$$\alpha p^{-\operatorname{ord} \alpha} = \begin{cases} \text{a quadratic residue (mod } p) & \text{if } p \text{ is odd,} \\ 1 \pmod 8, & \text{if } p = 2. \end{cases}$$

The converse is also true and is proved using the Hensel-Newton method (D). Hence $[\mathbf{Q}_p^\times : \mathbf{Q}_p^2] \leqq 8$. One can show that $[\mathbf{Q}_p^\times : \mathcal{S}] = 2$, by examining which of the coset leaders of $\mathbf{Q}_p^2$ lie in $\mathcal{S}$. It then follows that for all p (finite or infinite),

$$(a, b)_p(a, c)_p = (a, bc)_p, \quad \text{for all } a, b, c \in \mathbf{Q}_p.$$

For each finite prime, the congruence $x^2 - ay^2 - bz^2 \equiv 0 \pmod p$ is always solvable, either because $p|ab$ or because of Chevalley's theorem. Thus if p does not divide $2ab$, then by the Hensel-Newton method we can solve $x^2 - ay^2 - bz^2 = 0$ in $\mathbf{Q}_p$

and hence there is a solution with $x = 1$; i.e., $(a, b)_p = 1$. Thus

$$\prod_{\text{all } p} (a, b)_p = (a, b)_\infty \prod_{p|2ab} (a, b)_p.$$

Now by examining $(-1, -1)_p$, $(-1, p)_p$, $(-1,q)_p$ for various primes p and q and by using the above properties of the map $(a, b)_p$, it can be shown that

$$\prod_{\text{all } p} (a, b)_p = 1, \quad \text{for all } a, b \in \mathbf{Q}^\times.$$

Consequently, *if* $a, b \in \mathbf{Q}^\times$, *then the equation* $x^2 - ay^2 - bz^2 = 0$ *has a solution in all but finitely many local fields and the number of these exceptions is even.*

The proof of (J) for $m \geqq 4$ is now obtained by induction on m. Suppose (J) is true for forms with fewer than m variables and that $m \geqq 4$. The existence of real zeros allows us to assume $a_m < 0$. The existence of p-adic zeros implies the existence of a p-adic integer m_p such that $p^2 \nmid m_p$ and the equations

$$x_1^2 + a_2x_2^2 = m_p, \; a_3x_3^2 + \cdots + a_mx_m^2 = -m_p$$

are solvable in $\mathbf{Q}_p$. Let $m_p = p^{\lambda_p}u_p$, where u_p is a p-adic unit and $\lambda_p = 0$ or 1. By Dirichlet's theorem on primes in an arithmetic progression there is a prime

$$q \equiv \begin{cases} m_p/\Pi p^{\lambda_p} \pmod{p}, & \text{if } p \text{ is an odd prime } p|a_2\cdots a_m, \\ m_2/\Pi p^{\lambda_p} \pmod{16}, & \text{if } p = 2. \end{cases}$$

Set $d = q \prod_{p|2a_2\cdots a_m} p^{\lambda_p}$. Using the Hensel-Newton method one can show that the equation

$$x_1^2 + a_2x_2^2 = dz^2 \tag{10.1}$$

is solvable in all local fields, except possibly in $\mathbf{Q}_q$. But the number of local fields in which (10.1) has no solution must be even. Hence (10.1) is solvable in all local fields and hence in $\mathbf{Q}$. If $z \neq 0$, there is a solution with $z = 1$. If the solution has $z = 0$, then $a_2 = -e^2$, and again there would be a solution with $z = 1$.

Similarly,

$$a_3x_3^2 + \cdots + a_mx_m^2 = -dz^2 \tag{10.2}$$

is solvable in all local fields except possibly $\mathbf{Q}_q$. If $m = 4$, we can show, as above, that (10.2) has a rational solution with $z = 1$. Next suppose that $m \geqq 5$. Since $(q, 2a_2 \cdots a_m) = 1$, by the theorem of Chevalley, the congruence $\sum_{\nu=3}^{m} a_\nu x_\nu^2 \equiv 0 \pmod{q}$ is solvable with $2a_\nu x_\nu \not\equiv 0 \pmod{q}$, for some ν. Hence by the Hensel-Newton method, $\sum_{\nu=3}^{m} a_\nu x_\nu^2$ has a zero in $\mathbf{Q}_q$. But then (10.2) is equivalent over $\mathbf{Q}_q$ to a quadratic form in which one of the variables $x_3, \cdots, x_m$ appears only to the first degree, whence (10.2) has a solution in each local field, it must (by the induction hypothesis) have a rational zero. Knowing that (10.2) has a rational zero we can then show that it has one with $z = 1$. Combining this conclusion with the preceding paragraph, we see that $x_1^2 + a_2x_2^2 + \cdots + a_mx_m^2$ has a rational zero. This completes the proof of (J).

Using the theorem of Chevalley and the Hensel-Newton method, it is a relatively simple problem to prove: *A quadratic form in $m \geqq 5$ variables, with coefficients in $\mathbf{Z}$, has a zero in each $\mathbf{Q}_p$, where p is a rational prime.*

Combining this with (J) we have:

K. *An indefinite quadratic form in $m \geqq 5$ variables, with coefficients in $\mathbf{Z}$, has a zero in $\mathbf{Z}$.*

Our local conditions only tell us that a quadratic form F in m variables, with coefficients in $\mathbf{Z}$, has an integral zero. These conditions do not tell us how to find such a zero. However, Cassels [8] has shown that if a form F has a zero it must have one for which the absolute value of each of its coordinates is less than $[\frac{1}{2}(3m^2 + m - 10)(m^2 - 1)|F|]^{(m-1)/2}$, where $|F|$ = maximum of the absolute values of the coefficients of F. Hence, if we know that F has an integral zero, we can find one by trial and error in a finite length of time.

11. SEVERAL QUADRATIC FORMS

Conditions for the existence of rational zeros of a set of two or more quadratic forms are less fully determined. We do not know whether the existence of a zero in each local field implies the existence of a rational zero. Swinnerton-Dyer [32] has proved:

If F and G are quadratic forms over $\mathbf{Q}$ *in* $m \geqq 11$ *variables and each form in the pencil* $aF + bG$ *is indefinite, then the set* $\{F, G\}$ *has a rational zero.*

Clearly the indefinite condition is necessary. Since

$$F(x, y, z, w) = x^2 + y^2 - 3w^2 - 3z^2$$

has no rational zero, it is clear from $\{F(x, y, z, w), F(x', y', z', w')\}$ that nine variables are needed. Demyanov [18] (for a simpler proof see [5]) has shown that two quadratic forms over $\mathbf{Q}$ in $m \geqq 9$ variables have a zero in each p-adic field. Thus one might hope that one could replace the 11 in Swinnerton-Dyer's theorem by 9, but it appears that one will need a major new idea to achieve this hope. The idea behind Swinnerton-Dyer's proof is that to each pair $\{F, G\}$ there are infinitely many pairs $\{F^*, G^*\}$, where F^* has the shape $F^* = x_1x_2 + \cdots + x_7x_8 + H(x_9, x_{10}, x_{11})$, which are equivalent to $\{F, G\}$ over $\mathbf{Q}$. If $K^* = G^*(x_1, 0, x_3, 0, x_5, 0, x_7, 0, \cdots, 0)$ should have a zero in $\mathbf{Q}$, we would then have a rational zero of $\{F^*, G^*\}$, and hence a rational zero of $\{F, G\}$. One then characterizes those pairs F, G such that all the associated K^* fail to have a zero in $\mathbf{Q}$, and shows that such pairs F, G have a rational zero for other reasons. We omit the proof of this theorem as it entails some very ingenious and difficult arguments and makes considerable use of phenomena of local fields which we have not developed.

12. INTEGRAL ZEROS OF QUADRATIC POLYNOMIALS

Up to now we have concentrated our attention on the zeros of a quadratic form. Actually it is valuable and desirable to enlarge the problem to consider the set of values assumed by the quadratic form as the variables range over $\mathbf{Z}$, $\mathbf{Q}$, and $\mathbf{Q}_p$. Such an extension comes more to the fore when we view a quadratic form as an inner product on projective space. This point of view has been very productive and a rather complete theory has been developed.

(See Eichler [19], O'Meara [30].) Here again the local theory plays an important role.

If a quadratic form over **Z** has a rational zero it has such a zero with all its coordinates nonzero. It follows that if a quadratic form F over **Z** is such that the equation $F = a$ is soluble in each local field, then $F - ax_0^2$ has a zero in **Q** with $x_0 \neq 0$ and hence $F = a$ has a zero in **Q**. However, if $n \in \mathbf{Z}$ and $F = n$ is soluble in each local ring of integers, then $F = n$ need not be soluble in **Z**; e.g., $x^2 + 55y^2 = 5$ is soluble in each $\mathfrak{O}_p$ but not in **Z**. On the other hand, it has been shown that if F is an indefinite quadratic form over **Z** in $m \geqq 4$ variables and if $n \in \mathbf{Z}$, then $F = n$ is soluble in **Z** if and only if it is soluble in each $\mathfrak{O}_p$. Furthermore if for each prime p, the coordinates of the solution in $\mathfrak{O}_p$ are such that x_ν is in $a_\nu + nq\mathfrak{O}_p$, then there is a solution in **Z** with $x_\nu \in a_\nu + nq\mathbf{Z}$. Using this last result, G. L. Watson [37] proved:

If $f(x_1, \cdots, x_m) = Q + L + c$ *is a quadratic polynomial over* **Z** *in* $m \geqq 4$ *variables, where* $F = Q + x_0L + cx_0^2$ *is a nondegenerate form and* **Q** *is an indefinite form, then* f *has a zero in* **Z** *if and only if* f *has a zero in each local ring* $\mathfrak{O}_p$.

Outline of proof: If Q is equivalent to a form in fewer than m variables, f will always have an integral zero. We may, therefore, suppose that Q is an indefinite, nondegenerate form in m variables. One can choose integers $q, a_1, \cdots, a_m$ and n such that

$$q^2f(x_1, \cdots, x_m) = Q(qx_1 + a_1, \cdots, qx_m + a_m) - n.$$

One then shows that if f has p-adic integer zeros, then there exist $z_\nu \in \mathfrak{O}_p$ such that $Q(z_1, \cdots, z_m) = n$ and $z_\nu \in a_\nu + qn\mathfrak{O}_p$. Hence there are rational integer solutions of $Q(z_1, \cdots, z_m) = n$ with $z_\nu \in a_\nu + qn\mathbf{Z}$. These solutions induce integral zeros of f.

The assumption that F be nondegenerate is necessary, for the polynomial $f = \frac{1}{5}[(5x_1 + 1)^2 + (5x_2 + 1)^2 - 3(5x_3 + 2)^2 - 3(5x_4 + 2)^2]$ has zeros in each $\mathfrak{O}_p$, but has no zero in **Z**. When Q is definite, and when $m = 2$ or 3, one needs additional assumptions to imply the existence of an integral zero.

13. POLYNOMIALS OF HIGHER DEGREE

Unlike quadratic forms, which have been studied for over a century, it is only in recent decades that significant results have been obtained for general polynomials and forms of higher degree. R. Brauer [7] was the first to obtain a general result for forms of high degree. Birch [3] modified and extended Brauer's method to prove:

Let $d_1, \cdots, d_r$ be odd integers and k be a positive integer. There is an integer $M(d_1, \cdots, d_r; k)$, depending on the d_i and k, such that if $F_1, \cdots, F_r$ are forms over $\mathbf{Q}$ of degrees $d_1, \cdots, d_r$, respectively, in $m \geqq M(d_1, \cdots, d_r; k)$ variables, then the set $\{F_1, \cdots, F_r\}$ has a linear space of rational zeros of dimension k over $\mathbf{Q}$.

The proofs by Brauer and Birch are beautiful examples of the use of induction. It is essential in Birch's proof to know that for each odd integer d there is an integer $N(d)$ such that diagonal forms (those of the type $\sum_{\nu=1}^{m} a_\nu x_\nu^d$) with $m \geqq N(d)$ have zeros in $\mathbf{Q}$. The existence of the $N(d)$ has been known for several decades. Using the Hardy-Littlewood-Vinogradov circle method, Davenport and Lewis [14] have shown that an indefinite diagonal form over $\mathbf{Q}$ has a rational zero when $m \geqq d^2 + 1$ and $d \neq 7, 8, \cdots, 17$. A somewhat larger value than $d^2 + 1$ may be needed for these exceptional d. The circle method is too involved to be given here, and therefore we have also omitted the proof of the theorem of Birch. For a careful detailed proof see Davenport [13].

The Brauer-Birch method is extremely wasteful and undoubtedly their bound on the number of variables needed to imply a rational zero is unnecessarily large; e.g., they need over 500 variables to infer that a cubic form has a rational zero. It is therefore natural to seek other methods and proofs which require fewer variables. On the other hand, it should be noted that the existence of such a bound is a special property of forms of odd degree. The form $\left(\sum_{\nu=1}^{\mu} x_\nu^2\right)^2 - 3\left(\sum_{\nu=1}^{\mu} y_\nu^2\right)^2$ has no rational zero regardless of the size of μ, although it is indefinite and has p-adic zeros, for every p, when $\mu \geqq 5$.

In his paper, Brauer proved that there exists a function φ_p such that every form of degree d in $m \geqq \varphi_p(d)$ variables with coefficients in a p-adic field has a zero in that field. It was conjectured by Artin that one could take $\varphi_p(d) = d^2 + 1$. Such is indeed the case for $d \leqq 3$, but the conjecture is false for infinitely many d.

The first counter-example was given by Terjanian [35]. He observed that

$$\begin{aligned} G(x) &\equiv G(x_1, x_2, x_3) \\ &= x_1^4 + x_2^4 + x_3^4 - x_1^2x_2^2 - x_1^2x_3^2 - x_2^2x_3^2 \\ &\qquad - (x_1x_2x_3)(x_1 + x_2 + x_3) \end{aligned}$$

is such that

$$G(x) \equiv \begin{cases} 1 \pmod 4), & \text{if some } x_i \text{ is odd,} \\ 0 \pmod{16}, & \text{if all } x_i \text{ are even.} \end{cases}$$

Hence

$$F = G(x) + G(y) + G(z) + 4G(u) + 4G(v) + 4G(w)$$

is a form in $18(> 4^2 + 1)$ variables which has only the trivial solution in $\mathbf{Q}_2$.

Shortly after Terjanian found his example, a number of mathematicians showed that for each prime p there exists a form F_p (with rational integer coefficients) of degree $d = p^\alpha(p-1)$ in $n > d^2 + 1$ variables having only the trivial solution in $\mathbf{Q}_p$. Browkin [6] demonstrated even more. He showed the existence of a function $\lambda(p)$ such that $\lim_{p\to\infty} \lambda(p) = 0$ and the existence of forms G_p over $\mathbf{Z}$ of degree d in $n > d^{3-\lambda(p)}$ variables having only the trival solution in $\mathbf{Q}_p$. At the present time we have no examples to show that we need to take $\varphi_p(d) > d^3$; on the other hand, neither is anyone conjecturing that we can take $\varphi_p(d) = d^3$. There are reasons to believe that Artin's conjecture is valid when the degree is a prime — but no proof exists.

By means of ultraproducts and some sophisticated extensions of valuation theory, Ax and Kochen [1] proved: *There exists a function $P(d)$ such that a form over $\mathbf{Z}$ of degree d in $n \geqq d^2 + 1$ variables has a zero in $\mathbf{Q}_p$, provided $p > P(d)$.* It would follow that if $p > P(d)$, then $\varphi_p(d)$ can be taken to be $d^2 + 1$. Unfortunately, their proof does not enable one to estimate the size of $P(d)$, for any d. Paul Cohen has given another (so far unpublished) proof of the Ax-Kochen result. Both the original proof and Cohen's proof are long, and require the use of quite deep theorems from logic, and consequently we omit giving a proof here.

We next prove:

L. *A cubic form in* $m \geqq 10$ *variables, with coefficients in a p-adic field* $\mathbf{Q}_p$, *has a zero in* $\mathbf{Q}_p$.

Proof: Let $F = \sum_{i \leq j \leq k} c_{ijk}x_ix_jx_k$, where the $c_{ijk} \in \mathbf{Q}_p$. Let $\mathcal{G} = \mathcal{G}(F)$ be the resultant of the partial derivatives $\partial F/\partial x_1, \cdots, \partial F/\partial x_m$. Then $\mathcal{G}$ is a nonzero form of degree $m2^{m-1}$ in the coefficients of F, each term of $\mathcal{G}$ being of degree 2^{m-1} in the coefficients of each of the forms $\partial F/\partial x_j$. Consequently, if $(x_1, \cdots, x_m) = (y_1, \cdots, y_m)T$ and $F^*(y_1, \cdots, y_m) = aF(x_1, \cdots, x_m)$, then

$$\mathcal{G}(F^*) = a^{m2^{m-1}}(\det T)^{3\cdot 2^{m-1}}\cdot\mathcal{G}(F).$$

Since $\mathcal{G}$ is a nonzero form it cannot vanish on any neighborhood of a point.† Hence, if $\mathcal{G}(F) = 0$, then for each integer n there exist $c_{ijk}^{(n)} \in \mathbf{Q}_p$ such that $||c_{ijk} - c_{ijk}^{(n)}||_p < 1/n$ and $\mathcal{G}(F^{(n)}) \neq 0$. Suppose $(a_{1n}, \cdots, a_{mn})$ is a zero of $F^{(n)}$ in $\mathfrak{O}_p^{(m)}$. Since some $a_{in} \neq 0$, we can assume some a_{in} is a unit. By the compactness of $\mathfrak{O}_p^{(m)}$, these points have at least one accumulation point in $\mathfrak{O}_p^{(m)}$ which has at least one coordinate a unit of $\mathfrak{O}_p$. Since F is a continuous function, this accumulation point is a zero of F. Therefore we may restrict our attention to cubic forms F with $\mathcal{G}(F) \neq 0$.

If F and F^* are equivalent over $\mathbf{Q}_p$, then $\mathcal{G}(F) = 0$ if and only if $\mathcal{G}(F^*) = 0$. Every form over $\mathbf{Q}_p$ is equivalent to a form with

---

† This is seen by induction on the number of variables in the polynomial $\mathcal{G}$. It is well known for one variable. Suppose $\mathcal{G}$ vanishes on a neighborhood. Choose any point in the neighborhood, say $a = (a_1, \cdots, a_N)$. The neighborhood contains a line segment through a parallel to the hyperplane $c_1 = 0$. Hence $\mathcal{G}(c_1, a_2, \cdots, a_N) = 0$ vanishes on an interval and hence the polynomial coefficients $F_\nu(c_2, \cdots, c_N)$ of c_1^ν vanish at a. But a was arbitrary, hence each F_ν vanishes on the neighborhood. By induction, each is the zero polynomial and therefore $\mathcal{G}$ is the zero polynomial.

coefficients in $\mathfrak{O}_p$. If F has coefficients in $\mathfrak{O}_p$, then ord $\mathfrak{g}(F) = I(F) \geqq 0$. A form F, with coefficients in $\mathfrak{O}_p$, is said to be *reduced* if $\mathfrak{g}(F) \neq 0$ and $I(F) \leqq I(F^*)$ for all F^*, with coefficients in $\mathfrak{O}_p$, which are equivalent to F over $\mathfrak{O}_p$. By the well-ordering principle, every form with $\mathfrak{g}(F) \neq 0$ is equivalent to a reduced form. Hence we may restrict our attention to reduced forms.

To each cubic form F over $\mathfrak{O}_p$ there is a unique minimal integer r such that $F(x_1, \cdots, x_m) \equiv F_1(L_1, \cdots, L_r) \pmod{\mathfrak{p}}$, where F_1 is a form in r variables over $\mathfrak{O}_p$ and the L_i are linear forms with coefficients in $\mathfrak{O}_p$ which are linearly independent modulo $\mathfrak{p}$. Clearly, $r \leqq m$. We may make a unimodular substitution having $y_i = L_i (i = 1, \cdots, r)$ to obtain an equivalent form F^*, where $F^*(y_1, \cdots, y_m) \equiv F_1(y_1, \cdots, y_r) \pmod{\mathfrak{p}}$. If F is reduced, so is F^*. Let

$$F'(z_1, \cdots, z_m) = p^{-1}F^*(pz_1, \cdots, pz_r, z_{r+1}, \cdots, z_m),$$

Then F' has coefficients in $\mathfrak{O}_p$ and $I(F') = 2^{m-1}[3r - m] + I(F^*)$. Hence, if F is reduced, then $r \geqq m/3$.

Suppose F is reduced and $m \geqq 10$. Then $r \geqq 4$ and by the theorem of Chevalley, there exist integers $b_1, \cdots, b_r$, with some $b_i = 1$, such that $F_1(b_1, \cdots, b_r) \equiv 0 \pmod{\mathfrak{p}}$. Say $b_1 = 1$. Then $F^*(z_1, b_2z_1 + z_2, \cdots, b_rz_1 + z_r, z_{r+1}, \cdots, z_m) \equiv z_1^2L + z_1Q + C \pmod{\mathfrak{p}}$, where L, Q, and C are forms in $z_2, \cdots, z_r$. Since r is minimal, it follows that L and Q are not both identically $0 \pmod{\mathfrak{p}}$. If L is not identically $0 \pmod{\mathfrak{p}}$, then $1, 0, \cdots, 0$ is a solution of $F^* \equiv 0 \pmod{\mathfrak{p}}$ and some partial derivative of F^* does not vanish $\pmod{\mathfrak{p}}$ at $1, 0, \cdots, 0$. If L is identically $0 \pmod{\mathfrak{p}}$, we can choose $d_2, \cdots, d_m \in \mathbf{Z}$, not all $0 \pmod{\mathfrak{p}}$, so that $Q(d_2, \cdots, d_r) \not\equiv 0 \pmod{\mathfrak{p}}$. Then $-C(d_2, \cdots, d_r), d_2Q(d_2, \cdots, d_r), \cdots, d_mQ(d_2, \cdots, d_r)$ is a solution of $F^* \equiv 0 \pmod{\mathfrak{p}}$ with $\partial F^*/\partial x_1 \not\equiv 0 \pmod{\mathfrak{p}}$. In either case, we can now apply the Hensel-Newton method to obtain a zero of F^* and hence of F, which has coordinates in $\mathfrak{O}_p$. This completes the proof.

This theorem was first proved by Demyanov [17] and by Lewis [21]. The proof given here is related to, but differs from either of these earlier proofs.

Since a cubic form is always indefinite and since indefinite forms over **R** always have real zeros, it follows that: A cubic form over **Q** in $m \geqq 10$ variables has a zero in each local field of **Q**. However, as with the case of two quadratic forms, we have been unable to prove that for a cubic form the existence of local zeros implies the existence of a rational zero. H. Davenport [11] has proved: *A cubic form over* **Q** *in* $m \geq 16$ *variables has a zero in* **Q**.

In proving this result, Davenport had to develop and extend the circle method to polynomials of nonadditive type. (A polynomial is of additive type if it is of the form $\sum_{i=1}^{n} a_i x_i^{s_i}$.) This development was a major breakthrough and is much too complicated and involved to be given here. The case $m \geqq 17$ is somewhat simpler, but still long. An excellent exposition of that result can be found in [13]. In Davenport's proof, sixteen variables were needed because in the proof a certain series must converge and it is known to converge provided $m \geqq 16$. No counter-example exists to the conjecture that a cubic form in 10 variables has a rational zero.

In applying the circle method it is essential to know not only that the form has a zero in each local field but that the set of zeros in each local field contains a manifold of dimension $m - 1$. In the case of a cubic form, the existence of a single zero implies the existence of a nonsingular zero; i.e., one for which one of the partial derivatives does not vanish. As in the calculus, a nonsingular zero implies that the set of zeros contains a manifold of dimension $m - 1$. However, for forms of degree $\geqq 4$, a single zero need not imply the existence of a nonsingular zero. Hence to apply the circle method to forms of degree higher than 3 one needs at least to assume conditions which imply the existence of such a manifold of zeros in each local field. Even that is not enough, for Swinnerton-Dyer has pointed out that

$$G = 3(x_1^2 + \cdots + x_r^2)^3 + 4(x_{r+1}^2 + \cdots + x_s^2)^3 - 5(x_{s+1}^2 + \cdots + x_m^2)^3$$

has no rational zero. However, if r, $s - r$ and $m - s$ are sufficiently large, G does have nonsingular zeros in each local field and hence the desired manifolds. The form G is unusual in that the dimension of the set of zeros of $\{\partial G/\partial x_1, \cdots, \partial G/\partial x_m\}$ in the field of complex numbers is extremely large, namely $m - 3$. The dimension of this set of zeros is called the *singular dimension of* G. By an adaptation of Davenport's circle method, Birch [4] has shown that if we eliminate cases where the singular dimension is large, and if m is large, then the existence of the manifold of zeros of dimension $m - 1$ in each local field implies the existence of rational zeros. Specifically, Birch proved:

If a form F of degree d over $\mathbf{Q}$ *in m variables, with singular dimension s, has a nonsingular zero in each local field and if* $m - s \geqq (d - 1)2^d$, *then F has a zero in* $\mathbf{Q}$.

There are very few known conditions which imply the existence of integral zeros for nonhomogeneous polynomials. One of the few is the following [15, 38].

A cubic polynomial over $\mathbf{Z}$ *in* $m \geqq 20$ *variables having a zero in each local ring* $\mathfrak{O}_p$ *and such that its cubic part is not equivalent over* $\mathbf{Z}$ *to a form which has each of its terms divisible by one of* 3 *variables, has a zero in* $\mathbf{Z}$.

14. ALGEBRAIC NUMBER FIELDS†

An algebraic number field K is a finite algebraic extension of $\mathbf{Q}$. Each element of K is a zero of a polynomial with coefficients in $\mathbf{Z}$. A zero of a monic polynomial over $\mathbf{Z}$ is called an *algebraic integer*. The algebraic integers in K form an integral domain $\mathfrak{O} = \mathfrak{O}_K$, whose field of quotients is K. The domain $\mathfrak{O}$ has infinitely many proper prime ideals. Each of these prime ideals contains exactly one rational prime p. A nonzero element of $\mathfrak{O}$ (in particular, any rational prime) appears in only finitely many proper prime ideals of $\mathfrak{O}$. A prime ideal of $\mathfrak{O}$ need not be a principal ideal; however, each ideal of $\mathfrak{O}$ can be generated by two

† For proofs of statements made in this section see Artin [2], Hecke [20], or Pollard [31].

elements. Each prime ideal $\mathfrak{P}$ contains elements not in the ideal $\mathfrak{P}^2$. Such elements are called *prime elements of* $\mathfrak{O}$.

The roots of unity in K form a finite subgroup R of the group $\mathfrak{U}$ of units in the domain $\mathfrak{O}$. The group $\mathfrak{U}/R$ is a free abelian group with finitely many generators. If r_1 is the number of distinct isomorphisms of K into $\mathbf{R}$, and $2r_2$ is the number of distinct isomorphisms of K into $\mathbf{C}$, then $r = r_1 + r_2 - 1$ is the number of generators of $\mathfrak{U}/R$.

As in the rational case, we can associate with each proper prime ideal $\mathfrak{P}$ of $\mathfrak{O}$ the inverse limit of the rings $\mathfrak{O}/\mathfrak{P}^n$, $n = 1, 2, \cdots$. We denote this inverse limit ring by $\mathfrak{O}_{\mathfrak{P}}$. It is an integral domain of characteristic 0 which contains an isomorphic imbedding of the domain $\mathfrak{O}$. On $\mathfrak{O}_{\mathfrak{P}}$ we have the order function $\operatorname{ord}_{\mathfrak{P}} \alpha = \max \{n \in \mathbf{Z} | \alpha \in \mathfrak{P}^n\}$. The domain $\mathfrak{O}_{\mathfrak{P}}$ contains a unique maximal prime ideal consisting of the nonunits of $\mathfrak{O}_{\mathfrak{P}}$. We shall abuse our notation by also letting $\mathfrak{P}$ denote this prime ideal of $\mathfrak{O}_{\mathfrak{P}}$.

The order function permits us to define a nonarchimedean valuation, $\| \quad \|_{\mathfrak{P}}$, on $K_{\mathfrak{P}}$, the field of quotients of $\mathfrak{O}_{\mathfrak{P}}$. If p is the rational prime in $\mathfrak{P}$, then $\mathbf{Q}_p$ is imbedded in a natural fashion in $K_{\mathfrak{P}}$, and $\operatorname{ord}_{\mathfrak{P}} p = e_{\mathfrak{P}} = e \geqq 1$. Relative to the topology induced by the valuation, $K_{\mathfrak{P}}$ is cauchy complete and $\mathfrak{O}_{\mathfrak{P}}$ is both open and compact. The elements of $K_{\mathfrak{P}}$ are limits of series of the type: $\sum_{\nu=m}^{\infty} c_\nu \pi^\nu$, where $\pi \in \mathfrak{P}$, $\pi \notin \mathfrak{P}^2$ and $c_\nu \in J$, a fixed set of representatives of $\mathfrak{O}/_{\mathfrak{P}}$. $K_{\mathfrak{P}}$ is called the $\mathfrak{P}$-*adic field.*

Since $\| \quad \|_{\mathfrak{P}}$ is nonarchimedean, a series $\sum \gamma_\nu$, $\gamma_\nu \in K_{\mathfrak{P}}$, converges in $K_{\mathfrak{P}}$ if and only if $\{\gamma_\nu\}$ is a null sequence. Furthermore, if such a series has a limit, then every series obtained from it by rearranging and grouping of terms has the same limit.

15. POWER SERIES OVER $K_{\mathfrak{P}}$

If $f(x) = \sum_{\nu=0}^{\infty} \gamma_\nu x^\nu$, $\gamma_\nu \in K_{\mathfrak{P}}$, converges for all $x \in \mathfrak{O}_{\mathfrak{P}}$, then $\{\gamma_\nu\}$ is a null sequence. It follows that for any rational integer n, all but finitely many γ_ν are divisible in $\mathfrak{O}_{\mathfrak{P}}$ by π^n. Hence we can rearrange and group terms so that, for all $x \in \mathfrak{O}_{\mathfrak{P}}$, $f(x) = \sum_{\nu=m}^{\infty} A_\nu(x)\pi^\nu$, where $m \in \mathbf{Z}$ and the $A_\nu(x)$ are polynomials whose nonzero coefficients are units in $\mathfrak{O}_{\mathfrak{P}}$. Conversely, if $B_\nu(x)$ are polynomials over $\mathfrak{O}_{\mathfrak{P}}$, then $g(x) = \sum_{\nu=m}^{\infty} B_\nu(x)\pi^\nu$ converges for all $x \in \mathfrak{O}_{\mathfrak{P}}$ and we can rearrange and group terms so that $g(x) = \sum_{\nu=0}^{\infty} \beta_\nu x^\nu$, where $\{\beta_\nu\}$ is a null sequence in $K_{\mathfrak{P}}$.

As with the real numbers, a *nonzero power series with coefficients in $K_{\mathfrak{P}}$, which converges on a compact set, has only finitely many zeros in that set.* In particular a nonzero power series which converges for all $x \in \mathfrak{O}_{\mathfrak{P}}$ can have only finitely many zeros in $\mathfrak{O}_{\mathfrak{P}}$. Given the coefficients of such a power series, we can even give a bound on the number of zeros in $\mathfrak{O}_{\mathfrak{P}}$.

M. Theorem of Strassman: *If $f(x) = \sum_{\nu=0}^{\infty} \gamma_\nu x^\nu$, $\gamma_\nu \in K_{\mathfrak{P}}$, is a series which converges for all $x \in \mathfrak{O}_{\mathfrak{P}}$ and is nonzero for some $x \in \mathfrak{O}_{\mathfrak{P}}$, then $f(x)$ has at most $k = \max \{\nu | \operatorname{ord}_{\mathfrak{P}} \gamma_\nu$ is minimal$\}$ zeros in $\mathfrak{O}_{\mathfrak{P}}$.*

Proof: Since f is not the zero function, some $\gamma_\nu \neq 0$. Since f converges for all $x \in \mathfrak{O}_{\mathfrak{P}}$, $\{\gamma_\nu\}$ is a null sequence. Hence there is an α in $\mathfrak{O}_{\mathfrak{P}}$ such that each $\alpha\gamma_\nu \in \mathfrak{O}_{\mathfrak{P}}$. Consequently, since f and αf, $\alpha \neq 0$, have the same zeros, we may assume that the $\gamma_\nu \in \mathfrak{O}_{\mathfrak{P}}$ and some γ_ν is a $\mathfrak{P}$-adic unit. We then have

$$f(x) = F_0(x) + \pi F_1(x) + \pi^2 F_2(x) + \cdots,$$

where the F_ν are polynomials whose nonzero coefficients are $\mathfrak{P}$-adic units. By assumption F_0 is a nonzero polynomial of degree k. Apply the division algorithm to determine the sequence of polynomials, $G_1, H_1, G_2, H_2, \cdots$ such that

$$\begin{aligned}
&F_1 = G_1F_0 + H_1, && 0 \leqq \deg H_1 < k,\\
&F_2 - G_1H_1 = G_2F_0 + H_2, && 0 \leqq \deg H_2 < k,\\
&\vdots\\
&F_m - G_1H_{m-1} - \cdots - G_{m-1}H_1 = G_mF_0 + H_m, && 0 \leqq \deg H_m < k,\\
&\vdots
\end{aligned}$$

Since F_0 has a unit as leading coefficient and the F_ν have coefficients in $\mathfrak{O}_{\mathfrak{P}}$, it follows that the G_ν, H_ν have coefficients in $\mathfrak{O}_{\mathfrak{P}}$. Put $g(x) = 1 + \sum_{\nu=1}^{\infty} \pi^\nu G_\nu(x)$, $H(x) = F_0(x) + \sum_{\nu=1}^{\infty} \pi_1^\nu H_\nu(x)$. Then $H(x)$ is a polynomial of degree k with coefficients in $\mathfrak{O}_{\mathfrak{P}}$ and $G(x)$ is a power series which converges to a $\mathfrak{P}$-adic unit for all $x \in \mathfrak{O}_{\mathfrak{P}}$. Furthermore $f(x) = g(x)H(x)$. Consequently the zeros of $f(x)$ in $\mathfrak{O}_{\mathfrak{P}}$ are exactly the zeros of $H(x)$ in $\mathfrak{O}_{\mathfrak{P}}$ and they are at most k in number.

16. A WEIERSTRASS EXPANSION FOR CONTINUOUS FUNCTIONS

Let $\binom{x}{k}$ be the polynomial $\dfrac{x(x-1)\cdots(x-k+1)}{k!}$. For each $n \in \mathbf{Z}$ the value $\binom{n}{k}$ lies in $\mathbf{Z}$. Since $\binom{x}{k}$, being a polynomial, is a continuous function on $\mathfrak{O}_p$ and since $\mathbf{Z}$ is dense in $\mathfrak{O}_p$, it follows that $\binom{x}{k}$ is a map of $\mathfrak{O}_p$ into $\mathfrak{O}_p$. We also observe that $\binom{-1}{k} = (-1)^k$.

Let $\gamma_k \in \mathfrak{O}_{\mathfrak{P}}$. If $\{\gamma_k\}$ is a null sequence, then the series $\sum_{k=0}^{\infty} \gamma_k \binom{x}{k}$ converges in $\mathfrak{O}_{\mathfrak{P}}$ for all $x \in \mathfrak{O}_p$. Conversely, if $\sum_{k=0}^{\infty} \gamma_k \binom{x}{k}$ converges for all $x \in \mathfrak{O}_p$, then $\Sigma(-1)^k\gamma_k$ converges and hence $\{\gamma_k\}$ is a null sequence. Furthermore, if $\|\gamma_k\|_{\mathfrak{P}} < \epsilon$ for $k \geqq M$, then $\sum_{k=M}^{\infty} \gamma_k \binom{x}{k} \leqq \epsilon$, for all $x \in \mathfrak{O}_p$. Hence if the series $\sum_{k=0}^{\infty} \gamma_k \binom{x}{k}$ converges for $x \in \mathfrak{O}_p$, it converges uniformly for $x \in \mathfrak{O}_p$.

We associate with each continuous function f from $\mathfrak{O}_p$ into $\mathfrak{O}_{\mathfrak{P}}$ the sequence $\{\gamma_k\}$, defined by the relations

$$\gamma_0 = f(0),\ \gamma_k = f(k) - \sum_{j=0}^{k-1} \gamma_j \binom{k}{j}, \quad k = 1, 2, \cdots.$$

Clearly the $\gamma_k \in \mathfrak{O}_{\mathfrak{P}}$. The γ_k are also given by the relations

$$(16.1)\quad \gamma_k = f(k) - \binom{k}{1}f(k-1) + \binom{k}{2}f(k-2) - \cdots + (-1)^k\binom{k}{k}f(0).$$

We shall call the γ_k the *W-coefficients* of the continuous function f. Note that the γ_k are all 0 exactly when f is the zero function.

Next, following ideas of K. Mahler [26], we prove

N. *If $f(x)$ is a continuous function from $\mathfrak{O}_p$ to $\mathfrak{O}_{\mathfrak{P}}$ and $\{\gamma_k\}$ are its W-coefficients, then $\sum_{k=0}^{\infty} \gamma_k \binom{x}{k}$ converges uniformly on $\mathfrak{O}_p$ to $f(x)$. The function $f(x)$ is the zero function if and only if all $\gamma_k = 0$.*

Proof: Since f is continuous on the compact set $\mathfrak{O}_p$, it is uniformly continuous on $\mathfrak{O}_p$. Hence to each $s \in \mathbf{Z}$ there is an integer $T = T(s)$ such that

$$||f(x) - f(y)||_{\mathfrak{P}} < ||\pi||_{\mathfrak{P}}^{s}, \quad \text{whenever } ||x - y||_p < p^{-T}.$$

To each s, specify a complete set J_s of residues of $\mathfrak{O}$ modulo $\mathfrak{P}^s$. Let $g_s(x)$ be the function on $\mathfrak{O}_p$ having its values in J_s and such that

$$(16.2)\qquad ||f(x) - g_s(x)||_{\mathfrak{P}} < ||\pi||_{\mathfrak{P}}^{s}, \quad \text{for all } x \in \mathfrak{O}_p.$$

Then $g_s(x) = g_s(y)$, whenever $y - x \in \mathfrak{p}^T$. Since g_s is constant on each of finitely many open sets of the form $x + \mathfrak{p}^T$, $g_s(x)$ is a continuous function from $\mathfrak{O}_p$ to $\mathfrak{O}_{\mathfrak{P}}$.

Let ω be a primitive p^Tth root of unity. Then ω, ω^{-1}, and $(\omega^m - 1)/(\omega - 1) = \omega^{m-1} + \cdots + 1$, $m = 1, 2, \cdots$, are algebraic integers. We also have

$$\sum_{k=0}^{p^T-1} = \omega^{mk} \begin{cases} p^T, & \text{if } m \equiv 0 \pmod{p^T}, \\ 0, & \text{otherwise.} \end{cases}$$

Put

$$\lambda_n = p^{-T} \sum_{k=0}^{p^T-1} \omega^{-kn} g_s(k), \quad n = 0, 1, \cdots.$$

Since ω^{-1} and the $g_s(k)$ are algebraic integers, so are the $p^T\lambda_n$. Furthermore

$$\psi(h) = \sum_{n=0}^{p^T-1} \lambda_n \omega^{hn} = p^{-T} \sum_{k=0}^{p^T-1} \sum_{n=0}^{p^T-1} \omega^{(h-k)n} g_s(k) = g_s(h),$$
$$h = 0, 1, \cdots, p^T - 1.$$

If $h \equiv m \pmod{p^T}$, then $\psi(h) = \psi(m) = g_s(m) = g_s(h)$. Hence

$$g_s(x) = \sum_{n=0}^{p^T-1} \lambda_n \omega^{xn}, \quad \text{for all } x \in \mathbf{Z}.$$

Let $\gamma_{s0}, \gamma_{s1}, \cdots$ be the W-coefficients associated with g_s. Then

$$\begin{aligned}\gamma_{sn} &= \sum_{k=0}^{n} (-1)^k \binom{n}{k} g_s(n-k) \\ &= \sum_{k=0}^{n} (-1)^k \binom{n}{k} \sum_{u=0}^{p^T-1} \lambda_u \omega^{u(n-k)} \\ &= \sum_{u=0}^{p^T-1} \lambda_u \sum_{k=0}^{n} (-1)^k \binom{n}{k} \omega^{u(n-k)} \\ &= \sum_{u=0}^{p^T-1} \lambda_u (1-\omega^u)^n.\end{aligned}$$

Hence $p^T \gamma_{sn}(1-\omega)^{-n} = \sum_{u=0}^{p^T-1} p^T \lambda_u \left(\dfrac{1-\omega^u}{1-\omega}\right)^n$ are algebraic integers, and

$$(p^T\gamma_{sn})^{\phi(p^T)} \prod_{\substack{(k,p^T)=1 \\ 0<k<p^T}} (1-\omega^k)^{-n} = (p^T\gamma_{sn})^{\phi(p^T)}[\Phi_{p^T}(1)]^{-n}$$

is in $\mathfrak{O}$. Here $\Phi_{p^T}(x)$ is the cyclotomic polynomial of index p^T. Now $\Phi_{p^T}(x) = (x^{p^T}-1)/(x^{p^{T-1}}-1)$ and hence $\Phi_{p^T}(1) = p$. It then follows that

$$||\gamma_{sn}||_{\mathfrak{P}} \leqq ||p||_{\mathfrak{P}}^{(n/\phi(p^T))-T} = ||\pi||_{\mathfrak{P}}^{e([n/\phi(p^T)]-T)},$$

and hence $\{\gamma_{sn}\}$ is a null sequence. In particular, there is an integer $N(s)$ such that $||\gamma_{sn}||_{\mathfrak{P}} \leqq ||\pi||_{\mathfrak{P}}^s$, whenever $n \geqq N(s)$. It follows from (16.1) and (16.2) that

$$||\gamma_n - \gamma_{sn}||_{\mathfrak{P}} \leqq ||\pi||_{\mathfrak{P}}^s, \quad n = 0, 1, \cdots.$$

But then

$$\begin{aligned}||\gamma_n||_{\mathfrak{P}} &= ||(\gamma_n - \gamma_{sn}) + \gamma_{sn}||_{\mathfrak{P}} \leqq \max\{||\gamma_n - \gamma_{sn}||_{\mathfrak{P}}, ||\gamma_{sn}||_{\mathfrak{P}}\} \\ &\leq ||\pi||_{\mathfrak{P}}^s, \quad \text{provided } n \geqq N(s).\end{aligned}$$

As s was any positive integer, the sequence $\{\gamma_n\}$ is a null sequence.

Since $\{\gamma_n\}$ is a null sequence, the series $\sum_{n=0}^{\infty} \gamma_n \binom{x}{n}$ converges uniformly on $\mathfrak{O}_p$; say $g(x)$ is that limit. As the $\binom{x}{n}$ are polynomials, $g(x)$ is the limit of a sequence of continuous functions on the compact set $\mathfrak{O}_p$ and hence $g(x)$ is continuous on $\mathfrak{O}_p$. Furthermore, by the construction of the γ_n, $g(m) = f(m)$ for all $m \in \mathbf{Z}$. Since $\mathbf{Z}$ is dense in $\mathfrak{O}_p$, each $x \in \mathfrak{O}_p$ is the p-adic limit of a sequence of rational integers $\{m_k\}$. But then

$$g(x) = \lim g(m_k) = \lim f(m_k) = f(x),$$

and hence $f(x)$ and $g(x)$ are the same function. This completes the proof of (N).

We have proved that each continuous function from $\mathfrak{O}_p$ to $\mathfrak{O}_\mathfrak{P}$ is the uniform limit of a sequence of polynomials. It follows by a simple change of variable $x \to ax$ that each continuous function from a compact subset of $\mathbf{Q}_p$ to $K_\mathfrak{P}$ is the uniform limit of a sequence of polynomials. This result also holds for continuous functions from compact subsets of $\mathbf{R}$ into $\mathbf{C}$, and in those cases the result is known as Weierstrass' approximation theorem.

Combining Theorems M and N, we obtain:

If $f(x)$ is a continuous function from $\mathfrak{O}_p$ to $\mathfrak{O}_\mathfrak{P}$ and has infinitely many zeros in $\mathfrak{O}_p$, then f is the zero function and hence the W-coefficients of f are all 0.

Now $x\binom{x}{k} = (k+1)\binom{x}{k+1} + k\binom{x}{k}$. It therefore follows by induction that:

If $M = \min\{m, n\}$, then

$$(16.3) \quad \binom{x}{m}\cdot\binom{x}{n} = \sum_{k=0}^{M} \binom{M}{k}\binom{n+m-k}{M}\binom{x}{n+m-k}.$$

This rule permits us to quickly determine the W-coefficients of the product of two continuous functions in terms of the W-coefficients of each of the functions. We have

$$\gamma_n(fg) = \sum_{\substack{0 \le s \le n \\ 0 \le t \le n \\ s+t \ge n}} \binom{\min\{s,t\}}{s+t-n}\binom{n}{\min\{s,t\}}\gamma_s(f)\gamma_t(g).$$

Of course, $\gamma_n(f+g) = \gamma_n(f) + \gamma_n(g)$.

17. EXPONENTIAL EQUATIONS

In this section we show how the information gathered in the previous three sections may be used to obtain quantitative results on the zeros of certain diophantine equations. We shall prove:

O. *Let K be an algebraic number field. Let $A_1(x), \cdots, A_s(x)$ be nonzero polynomials over K and let $\beta_1, \cdots, \beta_s$ be nonzero elements of K, such that none of the ratios β_i/β_j, $i \neq j$, is a root of unity. Then*

$$f(x) = A_1(x)\beta_1^x + \cdots + A_s(x)\beta_s^x$$

has only finitely many zeros in $\mathbf{Z}$.

Proof: Let $\mathfrak{O}$ be the ring of integers of K, and let $\mathfrak{P}$ be a prime ideal of $\mathfrak{O}$ containing an odd prime p, such that the various β_j and the leading coefficients of the $A_j(x)$ are $\mathfrak{P}$-adic units. Let π be a prime element of $\mathfrak{P}$. Since the residue class field $\mathfrak{O}/\mathfrak{P}$ is finite, there is an integer M such that

$$\beta_j^M = 1 + \pi\alpha_j = 1 + \rho_j, \quad \alpha_j \in \mathfrak{O}_{\mathfrak{P}}, \quad \rho_j \in \mathfrak{P}; \qquad j = 1, \cdots, s.$$

Since $\{\rho_j^n\}$ is a null sequence, the series $1 + \sum_{n=1} \rho_j^n \binom{y}{n}$ converges to a continuous function $g_j(y)$ from $\mathfrak{O}_p$ to $\mathfrak{O}_{\mathfrak{P}}$. Furthermore

$$g_j(k) = (1 + \rho_j)^k = \beta_j^{kM}, \text{ for } k = 0, 1, \cdots.$$

If $y \in \mathfrak{O}_p$ then y is the p-adic limit of a sequence $\{k_n\}$ of rational integers. As usual, we define $(1 + \rho_j)^y = \lim_n (1 + \rho_j)^{k_n}$. Then $\beta_j^{My} = (1 + \rho_j)^y$ is a continuous function from $\mathfrak{O}_p$ to $\mathfrak{O}_{\mathfrak{P}}$. Since the continuous functions $g_j(y)$ and β_j^{My} agree on a dense set $\mathbf{Z}$ in $\mathfrak{O}_p$ they are the same function.

Put

$$f_r(y) = \sum_{j=1}^{s} A_j(My + r)\beta_j^{My+r} = \sum_{j=1}^{s} A_j(My + r)\beta_j^r g_j(y)$$

$$= \sum_{j=1}^{s} B_{rj}(y)g_j(y), \quad r = 0, 1, \cdots, M - 1.$$

Clearly the $f_r(y)$ are continuous functions from $\mathfrak{O}_p$ to $\mathfrak{O}_{\mathfrak{P}}$. A rational integer $x = My + r$ is a zero of $f(x)$ exactly when y is a zero of $f_r(y)$. If f has infinitely many zeros in $\mathbf{Z}$, then there is an r such that f_r has infinitely many zeros in $\mathbf{Z}$ and hence f_r is identically zero.

The polynomials $B_{rj}(y)$ and $A_j(x)$ have the same degree, and hence for each r not all the $B_{rj}(y)$ are zero. Let $t = \max_{r,j} \{\deg B_{rj}(y)\}$. Then

$$B_{rj}(y) = \sum_{n=0}^{t} \gamma_{r,j,n}\binom{y}{n}, \quad j = 1, \cdots, s, \quad r = 0, 1, \cdots, M-1,$$

and, for each r, not all the $\gamma_{r,j,n}$ are zero. Furthermore, it follows from (16.3) that

$$B_{rj}(y)g_j(y) = \sum_{k=0}^{\infty}\left(\sum_{\nu=0}^{\min\{t,k\}} \gamma_{r,j,\nu}\rho_j^{k-\nu}(1+\rho_j)^{\nu}\binom{k}{\nu}\right)\binom{y}{k}.$$

Hence

$$f_r(y) = \sum_{k=0}^{\infty}\left(\sum_{\nu=0}^{\min\{t,k\}}\binom{k}{\nu}\left[\sum_{j=1}^{s} \gamma_{r,j,\nu}\rho_j^{k-\nu}(1+\rho_j)^{\nu}\right]\right)\binom{y}{k}.$$

If $f_r(y)$ is identically zero, its W-coefficients must all vanish. We would thus obtain an infinite system of homogeneous linear equations having $\gamma_{r,1,0}, \cdots, \gamma_{r,s,t}$ as a nontrivial solution. Consequently, the determinant Δ of the coefficients of the first $s(t+1)$ of these equations must be 0. The entry in the Ith row and the $(sJ+T)$th column of Δ is $\binom{I}{J}\rho_T^{I-J}(1+\rho_T)^J$. Let Γ be the determinant whose I, $sJ+T$ entry is $\binom{I}{J}\rho_T^{I-J}$. Since $1+\rho_T$ is a $\mathfrak{p}$-adic unit, we see that $\Delta = 0$ if and only if $\Gamma = 0$.

If $\Gamma(N; \rho_1, \cdots, \rho_s)$ is an N by N determinant whose I, $sJ+T$ entry is $\binom{I}{J}\rho_T^{I-J}$, then

$$\Gamma(N; \rho_1, \cdots, \rho_s) = \prod_{j=2}^{s} (\rho_j - \rho_1)^{[(N-j)/s]}\,\Gamma(N-1; \rho_2, \cdots, \rho_s, \rho_1).\dagger$$

† A sketch of a proof. Let $\Gamma_0 = \Gamma(N; \rho_1, \cdots, \rho_s)$. For $k = 1, \cdots, N-1$ let Γ_k be obtained from Γ_{k-1} by adding $(-\rho_1)$ times the $(N-k)$th row to the $(N-k+1)$th. Let Γ_N be obtained from Γ_{N-1} by subtracting the

It would then follow that $\Gamma = \pm \prod_{s>t} (\rho_s - \rho_t)^u$, for some rational integer u. For Γ to be zero, we would need $\rho_j = \rho_k$, for some $j \neq k$; whence $\beta_j^m = 1 + \rho_j = 1 + \rho_k = \beta_k^M$. It follows therefore that $f_r(y)$ is identically zero only if some ratio $\beta_j/\beta_k, (j \neq k)$ is a root of unity contrary to hypothesis. This proves our theorem.

It is an immediate consequence of our argument that,

P. *If $f(x)$ is as in* (O) *and has infinitely many zeros in* $\mathbf{Z}$, *then except for finitely many zeros the others lie in finitely many arithmetic progressions with modulus M.*

It should be noted that if $\sigma_1, \cdots, \sigma_s$ are all the isomorphisms of an algebraic number field E into $\mathbf{C}$ and if $\beta_j = \beta^{\sigma_j}$, $A_j(x) = A^{\sigma_j}(x)$, where β and the coefficients of $A(x)$ lie in E, then $f(x)$ maps $\mathbf{Z}$ into $\mathbf{Z}$. It is most often in this form that one encounters such functions.

Given specific values for the β_j and for the coefficients of A_j one can determine the order of the coefficients of the power series expansion of the various f_r and then apply the theorem of Strassman to obtain an upper bound for the number of integral zeros of $f(x)$. For example, using such methods, it can be shown that

$$(1 + \sqrt{-7})^x - (1 - \sqrt{-7})^x - \sqrt{-7}\, 2^x$$

has at most three integral zeros. Actually $x = 1, 2$ are its only integral zeros.

---

first column from the kth column, for $k = 2, 3, \cdots, s_1 = \min\{s, N\}$. For $k = 2, \cdots, s_1$, the kth column of Γ_N has the factor $(\rho_k - \rho_1)$. Let Γ_{N+1} be the result of removing these factors, then $\Gamma_0 = \prod_{j=2}^{s_1} (\rho_j - \rho_1)\Gamma_{N+1}$. If $N \leqq s$, we are finished. If $N > s$, let Γ_{N+2} be obtained from Γ_{N+1} by subtracting the kth column from the $(s + k)$th, for $k = 2, 3, \cdots, s_2 = \min\{s, N - s\}$. Then the $(s + k)$th column of Γ_{N+2} has the factor $(\rho_k - \rho_1)$. Let Γ_{N+3} be the result of removing these factors. Then $\Gamma_{N+1} = \prod_{j=2}^{s_2} (\rho_j - \rho_1)\Gamma_{N+3}$. If $N \leqq 2s$, we are finished. If $N > 2s$, let Γ_{N+4} be obtained from Γ_{N+3} by subtracting the $(s + k)$th column from the $(2s + k)$th, for $k = 2, \cdots, s_3 = \min\{s, N - 2s\}$. Continuing this process a sufficient number of times yields the conclusion.

18. LINEAR RECURRENCES

Q. *Suppose* $f(x)$ *is as in* (O) *and that* $\deg A_j = n_j$. *Then the function* $f(x) = \sum_{j=1}^{s} A_j(x)\beta_j^x$ *is such that*

$$f(m+k) = \sum_{j=0}^{k-1} c_j f(m+j), \quad \textit{for all } m \in \mathbf{Z}, \tag{18.1}$$

where $G = x^k - c_{k-1} - \cdots - c_0$ *is the least common multiple of* $F_1^{n_1+1}, \cdots, F_s^{n_s+1}$ *and* F_j *is the unique minimal monic polynomial over* $\mathbf{Q}$ *having* β_j *as a zero.*

Proof: Let $\Delta A(x) = A(x) - A(x-1)$, and define Δ^t inductively; i.e., $\Delta^t A(x) = \Delta(\Delta^{t-1}A(x))$. Then

$$\Delta^t A = \sum_{j=0}^{t} \binom{t}{j}(-1)^j A(x-j),$$

and

$$A(x) - A(x-r) = \sum_{s=1}^{r} \binom{r}{s}(-1)^{s-1}\Delta^s A(x).$$

Clearly $\deg \Delta A < \deg A$, and hence $\Delta^{n+1}A = 0$ if $\deg A = n$.

Suppose that F is the minimal monic polynomial over $\mathbf{Q}$ having β as a zero, that $G = x^k - \sum c_{k-j}x^{k-j}$ is a polynomial over $\mathbf{Q}$ which is divisible by F^{n+1}, and that A is a polynomial with complex coefficients of degree n. Let $H(x, z) = z^k G(x/z)$. Put

$$L = A(x+k)\beta^k - \sum_{j=1}^{k} A(x+k-j)c_{k-j}\beta^{k-j}.$$

It is sufficient for the proof of (Q) to prove that $L = 0$. We have

$$\begin{aligned} 0 &= A(x+k)G(\beta) \\ &= L - \sum_{j=1}^{k} c_{k-j}[A(x+k) - A(x+k-j)]\beta^{k-j} \\ &= L - \sum_{j=1}^{k} c_{k-j}\beta^{k-j} \sum_{s=1}^{j} \binom{j}{s}(-1)^{s-1}\Delta^s A(x+k) \\ &= L + \sum_{s=1}^{k} (-1)^s \Delta^s A(x+k) \sum_{j=s}^{k} \binom{j}{s} c_{k-j}\beta^{k-j} \end{aligned}$$

$$= L + \sum_{s=1}^{k} (-1)^s \Delta^s A(x+k)(S!) \frac{\partial H}{\partial z^s}(\beta, 1)$$
$$= L,$$

since $\Delta^s A(x+k) = 0$ if $s > n$ and $(\partial H/\partial z^s)(\beta, 1) = 0$ if $s \leqq n$.

A sequence of algebraic numbers $\{f(n)\}$ satisfying a relation

$$f(m+h) = \sum_{j=0}^{h-1} c_j f(m+j), \quad c_j \in \mathbf{Q},\ h \geqq 1, \tag{18.2}$$

is called a *linear recurrence*. Such a sequence may satisfy many different linear relations of the form (18.2). Suppose

$$f(m+h) = \sum_{j=0}^{h-1} c_j f(m+j) = \sum_{j=0}^{h-1} d_j f(m+j).$$

Let $r = \max\{j | c_j \neq d_j\}$; then

$$f(m+r) = \sum_{j=0}^{r-1} \frac{c_j - d_j}{c_r - d_r} f(m+j),$$

and hence the recurrence satisfies a linear relation with a smaller h. Consequently, for a given recurrence, when h is minimal the coefficients $c_0, \cdots, c_{h-1}$ are unique. The minimal h is called the *order* of the recurrence and $G(x) = x^h - c_{h-1}x^{h-1} - \cdots - c_0$ is called the *companion polynomial* to the recurrence. Since h is minimal we must have $c_0 \neq 0$.

Let $\{f(n)\}$ be a linear recurrence of order h and having $G = x^h - \cdots - c_0$ as its companion polynomial. Form the formal power series $Y = \sum_{n=0}^{\infty} f(n)z^n$. Then there exist $b_0, \cdots, b_{n-1}$ so that

$$Y(1 - c_{h-1}z - \cdots - c_0 z^h) = b_0 + b_1 z + \cdots + b_{h-1}z^{h-1} = B(z).$$

Now $G = \prod_{i=1}^{T} G_i^{m_i}$; where the G_i are distinct monic irreducible polynomials over $\mathbf{Q}$. Since the G_i are irreducible, $G_1 \cdots G_T$ has distinct zeros in the complex field. Suppose that $\beta_{i1}, \cdots, \beta_{in_i}$ are the zeros of G_i. By partial fraction expansion, we find,

$$Y = \frac{B(z)}{z^h G(z^{-1})} = \frac{B(z)}{\prod_{i=1}^{T} \prod_{j=1}^{n_i} (1 - \beta_{ij}z)^{m_i}} = \sum_{i=1}^{T} \sum_{j=1}^{n_i} \sum_{r=1}^{m_i} \frac{\gamma_{ij}}{(1 - \beta_{ij}z)^r}.$$

Since the $f(n)$ and the β_{ij} are algebraic, so are the b_ν and the γ_{ij}. Finally, using the formal power series expansion for

$$\frac{1}{(1-\beta z)^{k+1}} = \sum_{n=0}^{\infty} A_k(n)(\beta z)^n,$$

where $A_k(x)$ is a polynomial of degree k over $\mathbf{Z}$, we see that

$$Y = \sum_{n=0}^{\infty} \sum_{i=1}^{T} \sum_{j=1}^{n_i} B_{ij}(n)\beta_{ij}^n z^n,$$

where each $B_{ij}(x)$ is a polynomial of degree $m_i - 1$, with algebraic numbers as coefficients. Thus the sequence $\{f(n)\}$ is exactly the sequence of values of the function $f(x) = \sum_{i,j} B_{ij}(x)\beta_{ij}^x$ at the nonnegative rational integers. We have

R. *The numbers in a linear recurrence are the values of the function* $f(x) = \sum A_j(x)\beta_j^x$ *at nonnegative integers, where* $\{\beta_j\}$ *are the distinct zeros of the companion polynomial* G, *the* A_j *are polynomials with algebraic numbers as coefficients and the degree of* A_j *is* 1 *less than the multiplicity of* $x - \beta_j$ *in* G.

We have thus established a correspondence between rational functions $B(z)/z^kG(z^{-1})$, linear recurrences, and the exponential functions $f(x) = \sum_{j=1}^{s} A_j(x)\beta_j^x$.

S. *Suppose* $f(x)$ *and* G *are as in* (Q), *then the linear recurrence* $\{f(n)\}$ *has* k *as its order and* G *as its companion polynomial.*

Proof: Suppose H is the companion polynomial for the recurrence $\{f(n)\}$ given by (R). Then

$$f(n) = \sum_{j=1}^{s} A_j(n)\beta_j^n = \sum_{j=1}^{t} B_j(n)\gamma_j^n$$

where $\{\gamma_j\}$ is the set of distinct zeros of H. Certain of the ratios $\{\gamma_2/\gamma_1, \cdots, \gamma_t/\gamma_{t-1}, \beta_1/\gamma_1, \cdots, \beta_s/\gamma_t\}$ may be roots of unity. Let M be the least common multiple of the orders of any roots of unity appearing in that set. There exists a minimal $r \geqq 0$ so that, after relabeling, no ratio of distinct elements in the set $\mathfrak{X} = \{\beta_1^M, \cdots, \beta_s^M, \gamma_1^M, \cdots, \gamma_r^M\}$ is a root of unity and for all

$i = 1, \cdots, t, \gamma_i^M \in \mathfrak{X}$. Combining terms, we see that there exist polynomials C_j such that the function

$$g(x) = \sum_{j=1}^{s} A_j(M_x)\beta_j^{Mx} - \sum_{j=1}^{t} B_j(M_x)\gamma_j^{Mx}$$

$$= \sum_{j=1}^{s} C_j(Mx)\beta_j^{Mx} - \sum_{i=1}^{r} C_{s+i}(Mx)\gamma_i^{Mx}$$

has infinitely many integral zeros. Because of the properties of the set $\mathfrak{X}$, (0) implies this is possible only if the polynomials C_j are identically zero. It follows that, after a relabeling, there is a partitioning of the set $\{\gamma_j\}$ into s disjoint, nonempty subsets $\{\gamma_1^{(1)}, \cdots, \gamma_{u_1}^{(1)}\}, \cdots, \{\gamma_1^{(s)}, \cdots, \gamma_{u_s}^{(s)}\}$ such that $(\gamma_i^{(j)})^M = \beta_j^M$. Since $C_j = A_j - \sum_{i=1}^{u_j} B_i^{(j)} = 0$, we see that to each j there is an $i = i(j)$ such that $\deg B_{i(j)}^{(j)} \geq \deg A_j$. Hence the multiplicity of $x - \gamma_{i(j)}^{(j)}$ in H is at least as great as the multiplicity of $x - \beta_j$ in G. Since this is true for each j and since the $\gamma_{i(j)}^{(j)}$ are all distinct, we have $\deg H \geq \deg G$. But H was the companion polynomial and hence $\deg H \leqq \deg G$. Hence $\deg H = \deg G$, and by the argument preceding the definition of the companion polynomial, we see that $H = G$.

T. *Suppose $\{f(n)\}$ is a linear recurrence such that none of the zeros and none of the ratios of the distinct zeros of its companion polynomial is a root of unity, and suppose that $D(x)$ is a polynomial with algebraic numbers as coefficients. Then $f(n) = D(n)$ for only finitely many nonnegative integers n. In particular, no algebraic number appears infinitely often in the sequence $\{f(n)\}$.*

Proof: We know that $f(n) = \Sigma A_j(n)\beta_j^n$, where the A_j are polynomials and the β_j are the zeros of the companion polynomial. The function $g(x) = f(x) - D(x)1^x$ satisfies the hypothesis of (O) and hence has only finitely many rational integral zeros.

By careful use of the theorem of Strassman, one can show that for each recurrence satisfying the hypothesis of (T) there is an integer N such that no algebraic number appears in the recurrence more than N times. There are reasons to believe that the N can be so chosen so as to depend only on the order of the linear recurrence. However, this has not been shown even for recurrences of order 2. It is conjectured that for recurrences of order 2, N should be 5. Mahler has shown that when the companion polynomial has no multiple zeros and none of its zeros is a root of unity, then $\lim_{n\to\infty} |f(n)| = \infty$. However, his methods do not tell us how fast $|f(n)|$ tends to infinity.

19. REPRESENTATION BY A CUBIC FORM

In this section we shall use our results on exponential equations to obtain bounds on the number of integral zeros of certain polynomial equations.

Let $C(x, y)$ be an irreducible binary cubic form, with rational integer coefficients, and such that $C(x, 1)$ has exactly one real zero, say θ. Let $K = \mathbf{Q}(\theta)$ and let $\mathfrak{O}_K$ be the ring of algebraic integers in K. Each $\alpha \in K$ can be expressed as $\alpha = x + y\theta + z\theta^2$, with $x, y, z \in \mathbf{Q}$. There are exactly three isomorphisms of K into $\mathbf{C}$. If $\alpha \in K$, let $\alpha, \alpha', \alpha''$ denote the images of α under these three isomorphisms. The norm of α is defined to be the product $\alpha\alpha'\alpha''$. Hence

$$\text{norm }(\alpha) = N(\alpha) = N(x + y\theta + z\theta^2) = C(x,y) + zG(x, y, z),$$

where G is a quadratic form with rational coefficients. Hence x, y is a rational solution of $C(x, y) = 1$ exactly when $x + y\theta$ has norm 1.

The units of $\mathfrak{O}_K$ are exactly those integers of $\mathfrak{O}_K$ with norm ± 1. It is shown in algebraic number theory that corresponding to θ there is a rational integer $D(\theta)$ such that $x + y\theta + z\theta^2$ is in $\mathfrak{O}_K$ if and only if the denominators of x, y and z, in lowest terms, are divisors of $D(\theta)$.

Since $K \subset \mathbf{R}$, the only roots of unity in K are ± 1. By Dirichlet's theorem on units, there is a unit ϵ in $\mathfrak{O}_K$, not a root of unity, such that each unit in $\mathfrak{O}_K$ is of the form $\pm\epsilon^n, n \in \mathbf{Z}$. Of course, $-\epsilon$ would also serve this same role and hence we can assume that $N(\epsilon) = 1$.

U. *There are only finitely many units of $\mathfrak{O}_K$ of the form $x + y\theta$, with $x, y \in \mathbf{Q}$.*

Proof: We do know that

$$\epsilon^n = x_n + y_n\theta + z_n\theta^2, \quad x_n, y_n, z_n \in \mathbf{Q}.$$

Hence we have

$$(\epsilon')^n = x_n + y_n\theta' + z_n(\theta')^2,$$
$$(\epsilon'')^n = x_n + y_n\theta'' + z_n(\theta'')^2.$$

On solving this system of linear equations, we find

$$z_n = \gamma\epsilon^n + \gamma'(\epsilon')^n + \gamma''(\epsilon'')^n,$$

where $\gamma, \gamma', \gamma'' \in E = \mathbf{Q}(\theta, \theta', \theta'')$. Since ϵ is not in $\mathbf{Q}$ and K is a cubic field, we have $K = Q(\epsilon)$, and $E = Q(\epsilon, \epsilon', \epsilon'')$.

If η is a seventh root of unity, then

$$z_n = \gamma\epsilon^n + \gamma'(\epsilon')^n + \gamma''(\epsilon'')^n,$$

where $\gamma, \gamma', \gamma'' \in E = \mathbf{Q}(\theta, \theta', \theta'')$. Since ϵ is not in $\mathbf{Q}$ and K is a cubic field, we have $K = \mathbf{Q}(\epsilon)$ and $E = \mathbf{Q}(\epsilon, \epsilon', \epsilon'')$. The numbers ϵ' and ϵ'' are not real, otherwise $\mathbf{Q}(\theta') \subset E \subseteq \mathbf{R}$.

Suppose $\epsilon'/\epsilon = \eta$ is a root of unity. Since ϵ' is not real and ϵ is real, η is a nonreal element of E. Hence $E = \mathbf{Q}(\theta, \eta) = \mathbf{Q}(\epsilon, \eta)$. Since the degree of E over $\mathbf{Q}$ is 6, η must satisfy an irreducible monic quadratic polynomial with coefficients in K; say

$$(x - \eta)(x - \eta^*) = x^2 - Ax + B.$$

Since some power of η is 1, the same is true for its conjugate η^*. Hence η^* is also a root of unity. It then follows that $\eta\eta^* = B$ is a real root of unity and hence $B = \pm 1$. Let $\eta = a + bi$, where a, b are real numbers. Since η is nonreal, $b \neq 0$, we have $\eta\eta^* = (a + bi)(a - bi) = a^2 + b^2 = B = \pm 1$. It follows that $B = 1$, $a = 0$, $b = \pm 1$, $\eta = \pm i$, $\eta^* = -\eta$. Thus $\epsilon' = \pm i\epsilon$ and ϵ' satisfies the equation $x^2 + \epsilon^2 = 0$. Since ϵ'/ϵ is a root of unity so are ϵ''/ϵ' and ϵ''/ϵ and so ϵ'' also satisfies this equation. Hence ϵ', ϵ'' are $i\epsilon, -i\epsilon$ in some order. But then the minimal equation over $\mathbf{Q}$ satisfied by ϵ would be

$$(x - \epsilon)(x - i\epsilon)(x + i\epsilon) = x^3 - \epsilon x^2 + \epsilon^2 x - \epsilon^3;$$

contrary to ϵ being nonrational.

It follows that ϵ'/ϵ cannot be a root of unity and neither are either of the ratios ϵ''/ϵ, ϵ'/ϵ''. It follows from (O) that

$$z_n = \gamma\epsilon^n + \gamma'(\epsilon')^n + \gamma''(\epsilon'')^n$$

has only finitely many integral zeros. Correspondingly, ϵ^n is of the form $x + y\theta$ only finitely often.

If $C(x, 1)$ is monic, then $\theta \in \mathfrak{O}_k$ and $x + y\theta + z\theta^2 \in \mathfrak{O}_k$ when $x, y, z \in \mathbf{Z}$. We thus have as a corollary,

V. *If $C(x, y)$ is an irreducible cubic form over $\mathbf{Z}$ such that $C(x, 1)$ is monic and has exactly one real zero, then $C(x, y) = 1$ has only finitely many integral solutions.*

It is true, although we shall not prove it, that to each $a \in \mathbf{Z}$ there are finitely many elements of $\mathfrak{O}_k$, say $\alpha_1, \cdots, \alpha_m$ such that $\alpha \in \mathfrak{O}_k$ and $N(\alpha) = a$ if and only if $\alpha = \alpha_j\epsilon^n$, for some j and some n. Using this fact with the method above, we can show

If $C(x, y)$ is as in (V) *and $a \in \mathbf{Z}$, then the equation $C(x, y) = a$ has only finitely many integral solutions.*

Delaunay [16] and Nagell [28], using other methods, independently proved the following:

If $C(x, y)$ is an irreducible binary cubic form, with integral coefficients which has exactly one real zero, then $C(x, y) = 1$ has at most five integral solutions; and except for forms of a very special type, $C(x, y) = 1$ has at most three integral solutions.

20. THE EQUATION $x^3 - dy^3 = 1$

We now show how the theorem of Strassman may be used to obtain more precise results. We shall prove:

W. *If d is a cube free rational integer, then the equation $x^3 - dy^3 = 1$ has at most one integral solution in addition to the trivial solution* 1, 0.

Proof: Since $(-d)(-y)^3 = dy^3$, we may assume that $d > 1$. The polynomial $C(x, 1) = x^3 - d$ is monic and irreducible over $\mathbf{Q}$, and has exactly one real zero: $\theta = \sqrt[3]{d}$. The other zeros are $\theta\omega$, $\theta\omega^2$, where $\omega = \frac{1}{2}(-1 + \sqrt{3}i)$.

Suppose x, y and x_1, y_1 are nontrivial solutions for the equation $x^3 - dy^3 = 1$. Then $(x, y) = 1 = (x_1, y_1)$, $(x, d) = 1 = (x_1, d)$, and $yy_1 \neq 0$. As we saw earlier,

$$\alpha = x + y\theta \quad \text{and} \quad \alpha_1 = x_1 + y_1\theta$$

are units in $K = \mathbf{Q}(\theta)$ with positive norm. Hence they are of the form $\pm\epsilon^k$, where ϵ is a unit in $\mathfrak{O}_k$ of norm 1. In a cubic field, norm $(-\beta) = -$norm β, and consequently the minus sign does not occur. It follows that there exist integers m, n such that

$$\alpha = x + y\theta = \epsilon^m, \quad \alpha_1 = x_1 + y_1\theta = \epsilon^n$$

and hence there are integers u, v such that

$$\alpha^u = \alpha_1^v, \quad (u, v) = 1.$$

[For example, take $u = n/(m, n)$, $v = m/(m, n)$.]

Since $d > 1$ and cube free, there are primes p such that $p|d$, $p^3 \nmid d$. We choose one such prime p and keep it fixed for the remainder of the argument. By interchanging the two solutions x, y and x_1, y_1, we can assume that $p \nmid v$. Also, since $p|d$, we must have $p \nmid xx_1$. Furthermore, $x^3 - d$ is irreducible over $\mathbf{Q}_p$. For, otherwise, $d = \beta^3$ for some β in $\mathbf{Q}_p$ and since $||d||_p < 1$, we have $||\beta||_p < 1$ and therefore would have $p^3|d$, contrary to hypothesis. Since $x^3 - d$ is irreducible over $\mathbf{Q}_p$, $\{1, \theta, \theta^2\}$ is a linearly independent set over $\mathbf{Q}_p$. Let $\mathfrak{P}$ be the prime ideal of $K = \mathbf{Q}(\theta)$ such that $K_{\mathfrak{P}} = \mathbf{Q}_p(\theta)$. Since $p|d$ and $d = \theta^3$, we have $||\theta||_{\mathfrak{P}} < 1$.

Now, as noted above, $p \nmid x$ and hence $\beta = y/x$ is a p-adic integer, and $||\beta\theta||_{\mathfrak{P}} < 1$. The function $g(t) = (1 + \beta\theta)^t$ has the W-expansion:

$$g(t) = \sum_{\nu=0}^{\infty} \binom{t}{\nu} (\beta\theta)^t,$$

which converges for all p-adic integers t. In particular, it converges when $t = z = u/v$, since $p \nmid v$ and hence $||z||_p \leqq 1$. We therefore have

$$\alpha_1 = \alpha^{u/v} = \alpha^z = x^z(1 + \beta\theta)^z = x^z g(z)$$

$$= x^z \sum_{j=0}^{\infty} \binom{z}{3j} \beta^{3j} d^j + x^z \sum_{j=0}^{\infty} \binom{z}{3j+1} \beta^{3j+1} d^j \theta$$

$$+ x^z \sum_{j=0}^{\infty} \binom{z}{3j+2} \beta^{3j+2} d^j \theta^2.$$

Since $\alpha_1 = x_1 + y_1\theta$, and since $\{1, \theta, \theta^2\}$ is a linearly independent set over $\mathbf{Q}_p$, we must have (note $\beta^2 x^z \neq 0$)

$$h(z) = \sum_{j=0}^{\infty} \binom{z}{3j+2} \beta^{3j} d^j = 0;$$

i.e., z is a p-adic integral zero of the function

$$h(t) = \binom{t}{2} + \binom{t}{5}\beta^3 d + \binom{t}{8}\beta^6 d^2 + \cdots.$$

If $d \neq 2$ or $4 \mid yd$ the coefficient of t^2 in $h(t)$ is congruent to $\frac{1}{2} \pmod{p}$, while the coefficients of $t^s (s > 2)$ are all divisible by p. It therefore follows from Strassman's theorem (M) that $h(t)$ has at most two p-adic integral zeros. Since $\binom{0}{n} = \binom{1}{n} = 0$ when $n \geqq 2$, we see that 0 and 1 are the only p-adic integral zeros of $h(t)$. Since $z = u/v$ is a p-adic integral zero of $h(t)$, we must have $u/v = 1$ or 0; but then $\alpha_1 = \alpha$, or $\alpha_1 = 1$ (contrary to α_1 being nontrivial). If $d = 2$ the only solutions are 1, 0 and $-1, -1$. Thus $x^3 - dy^3 = 1$ can have at most one nontrivial solution and this proves the theorem.

The use of p-adic power series to solve diophantine equations was first introduced by Th. Skolem [33, 34], and accordingly such methods are usually referred to as *Skolem's p-adic method.* There is considerable art needed in applying the method, since the prime p must be chosen appropriately.

Using other more powerful methods, Nagell [29] has actually found a stronger result. He proved: If $x^3 - dy^3 = 1$ *has a nontrival solution* x_1, y_1, *then* $x_1 + y_1\theta = \epsilon$ *or* ϵ^2, *and it is* ϵ^2 *only for three values of d.*

21. GENERALIZATIONS AND EXTENSIONS

Let K be an algebraic number field of degree n. Suppose $\theta_1, \cdots, \theta_t$ are algebraic integers of K which are linearly independent over $\mathbf{Q}$. The set

$$\mathfrak{M} = \{a_1\theta_1 + \cdots + a_t\theta_t | a_i \in \mathbf{Z}\}$$

is a module in $\mathfrak{O}_K$ of rank t. In the preceding sections we showed that, for certain cubic fields, the module $a + b\theta$ contained only finitely many units. Th. Skolem [34] has proved:

If K is of degree 5 and has but one real conjugate, then all rank 3 modules in $\mathfrak{O}_K$ contain only finitely many units of $\mathfrak{O}_K$.

C. Chabauty in his famous thesis [9] extended these results and proved:

If K is of prime degree n and has r_1 real conjugates and $2r_2$ complex conjugates and if $t \leqq n - r_1 - r_2$, then each rank t module of $\mathfrak{O}_K$ contains only finitely many units of $\mathfrak{O}_K$.

Actually, Chabauty proved a more general result of which the above is a corollary.

Using his p-adic method, Skolem [33] proved,

If $F(x, y)$ is an irreducible binary form of degree $m \geqq 3$ over $\mathbf{Z}$ such that $F(x, 1)$ has exactly one real zero, then, for each $a \in \mathbf{Z}$, $F(x, y) = a$ has only finitely many integral solutions.

This theorem had been proved previously, without any assumptions on the zeros of $F(x, 1)$ by Thue [36], using a weak form of what has become known as the Thue-Siegel-Roth theorem. The Thue-Siegel-Roth theorem is concerned with integral solutions of certain inequalities. Neither proof gives an estimate for the number of integral solutions of $F(x, y) = a$. However, Mahler [27] has given a p-adic version of the Thue-Siegel-Roth theorem, and on using this he did obtain bounds on the number of integral solutions in terms of m, the coefficients of F, and the number of prime factors in a. For the sharpest result to date see [22].

As in the case of $x^3 - dy^3 = 1$, given a specific equation, one can often obtain very sharp bounds on the number of its integral zeros. Among such results are the following due to W. Ljunggren [23, 24, 25],

Each of the equations $x^2 - 1 = dy^4$ and $x^4 - 1 = dy^2$ has at most two positive integral solutions. Each of the equations $ax^4 - by^4 = c$ ($c = \pm 1, \pm 2, \pm 4, \pm 8$) has at most one positive integral solution.

To obtain such exceptionally good bounds, one must be very adept in choosing the prime with which to work. In his work, Ljunggren often makes use of the concept of class number of a field.

In these last sections, we have treated problems which entailed, at any one time, the determination of the integral solutions of one exponential equation in one unknown. There are obvious extensions of this problem. For example, while a single exponential equation in several variables is quite likely to have infinitely many integral zeros, there are fairly obvious conditions on a system of two exponential equations in two unknowns which imply that the system has only finitely many integral solutions. Problems such as this have been treated by Skolem and Chabauty.

A. Baker has very recently proved the following remarkable result:

To each form $F(x, y)$ of degree $\geqq 3$ and to each integer a there is a box B in the xy-plane, which can be explicitly determined knowing f and m, so that all integral zeros x, y of $F(x, y) = a$ lie in the box B.

Thus in principle, given $F(x, y) = a$ we can in a finite amount of time find all its zeros. In actuality, in most instances this is not a practical proposition since B is usually very large. However, this is the first time anyone has given a general procedure for finding all zeros of a large class of polynomials. Baker's proof uses his recent results on approximating logarithms of algebraic numbers by rational numbers. The proof will appear in the *Phil. Trans. Royal Society, London.*

BIBLIOGRAPHY

1. Ax J., and Kochen, S., "Diophantine problems over local fields, I," *Amer. J. Math.*, 87 (1965), 605–630.
2. Artin, E., *Theory of Algebraic Numbers.* Göttingen, 1959.
3. Birch, B. J., "Homogeneous forms of odd degree in a large number of variables," *Mathematika*, 4 (1957), 102–105.

4. ———, "Forms in many variables," *Proc. Royal Soc. London A*, 265 (1962), 245–263.

5. ———, Lewis, D. J., and Murphy, T., "Simultaneous quadratic forms," *Amer. J. Math.*, 74 (1962), 110–115.

6. Browkin, J., "On forms over p-adic fields," *Bull. Acad. Polon. Sci., Sér. Sci. Math. Astronom. Phys.*, 14 (1966), 489–492.

7. Brauer, R., "A note on systems of homogeneous algebraic equations," *Bull. Amer. Math. Soc.*, 51 (1945), 749–755.

8. Cassels, J. W. S., "Bounds for the least solutions of homogeneous quadratic equations," *Proc. Cambridge Philos. Soc.*, 51 (1955), 262–264.

9. Chabauty, C., "Sur les équations diophantiennes liées aux unités d'un corps de nombres algébriques fini," *Annali di Math.*, IV, 17 (1938), 127–168.

10. Chevalley, C., "Démonstration d'une hypothèse de M. Artin," *Abh. Math. Sem. Univ. Hamburg*, 11 (1935), 73–75.

11. Davenport, H., "Cubic forms in 16 variables," *Proc. Royal Soc. London A*, 272 (1963), 285–303.

12. ———, *The Higher Arithmetic*. London, 1952.

13. ———, *Analytic methods for diophantine equations and diophantine inequalities*. Ann Arbor, 1962.

14. ———, and Lewis, D. J., "Homogeneous additive equations," *Proc. Royal Soc. London A*, 274 (1963), 443–460.

15. ———, and Lewis, D. J., "Nonhomogeneous cubic equations," *J. London Math. Soc.*, 39 (1964), 657–671.

16. Delaunay, B., "The number of representations of an integer by a cubic form with negative discriminant," *Bull. Acad. Sci. USSR* (1922) (Russian).

17. Demyanov, V. B., "On cubic forms in discretely normed fields," *C. R. (Doklady) Acad. Sci. USSR*, 74 (1950) (Russian).

18. ———, "Pairs of quadratic forms over a complete field with discrete norm with a finite residue class field," *Izv. Akad. Nauk USSR*, 20 (1956), 307–324 (Russian).

19. Eichler, M., *Quadratische Formen und orthogonale Gruppen*. Berlin, 1952.

20. Hecke, E., *Algebraische Zahlen.* Leipzig, 1923.

21. Lewis, D. J., "Cubic homogeneous polynomials over p-adic fields," *Ann. of Math.* (2), 56 (1952), 473–478.

22. ———, and Mahler, K., "On the representations of integers by binary forms," *Acta Arith.*, 6 (1961), 333–363.

23. Ljunggren, W., "Über die Lösung einiger unbestimmten Gleichungen vierten Grades," *Oslo Vid. Akad. Avh. I*, 14 (1934).

24. ———, "Über die Gleichung $x^4 - Dy^2 = 1$," *Archiv for Math. og Naturvidenskab*, (5), 65 (1942).

25. ———, "Zur Theorie der Gleichung $x^2 + 1 = Dy^4$," *Oslo Vid. Akad. Avh I*, 5 (1942).

26. Mahler, K., "An interpolation series for continuous functions of a p-adic variable," *J. Reine Angew. Math.*, 190 (1958), 23–24.

27. ———, *Lectures on Diophantine Approximations.* Notre Dame, 1961.

28. Nagell, T., "Darstellung ganzer Zahlen durch binäre kubische Formen mit negativer Diskriminante," *Math. Z.*, 28 (1928), 10–29.

29. ———, "Solution complète de quelques équations cubiques à deux indéterminées," *J. de Math.*, 9ᵉ sér., 4 (1925), 209–270.

30. O'Meara, O. T., *Introduction to Quadratic Forms.* New York, 1963.

31. Pollard, H., *The Theory of Algebraic Numbers.* New York, 1950.

32. Swinnerton-Dyer, H. P. F., "Rational zeros of two quadratic forms," *Acta Arith.*, 9 (1964), 261–270.

33. Skolem, Th., Ein Verfahren zur Behandlung gewisser exponentialer Gleichungen und diophantischer Gleichungen. *8th Skand. Mat. Kong. Stockholm*, 1934, 163–188.

34. ———, "Einige Sätze über p-adische Potenzreihen mit Anwendung auf gewisse exponentielle Gleichungen," *Math. Ann.*, III (1935), 399–424.

35. Terjanian, G., "Un contre-exemple à une conjecture d'Artin," *C. R. Acad. Sci. Paris, Sér. AB*, 262 (1966), A 612.

36. Thue, A., "Über Annäherungswerte algebraischen Zahlen," *J. Reine Angew. Math.*, 135 (1909), 284–305.

37. Watson, G. L., "Indefinite quadratic diophantine equations," *Mathematika,* 8 (1961), 32–38.

38. ———, "Nonhomogeneous cubic equations," *Proc. London Math. Soc.*, 17 (1967), 271–295.

DIOPHANTINE DECISION PROBLEMS

Julia Robinson

1. INTRODUCTION

In 1900, Hilbert asked for a general method to determine if an arbitrary diophantine equation has a solution. Since the work of Gödel in 1931 and the development of the theory of computability, logicians have modified Hilbert's problem. They ask instead: Is there such a method? This is the *decision problem* for diophantine equations. The answer is not known even today.

In general, the *decision problem* for a set of questions is to find out if there is a general method to answer the questions. In the case of diophantine equations, the questions are those of the form: Does the diophantine equation $P(x_1, \cdots, x_k) = 0$ have a solution? Here P is a polynomial in $x_1, \cdots, x_k$ with integer coefficients. If we consider the wider class of exponential diophantine equations, i.e., equations in which variables may occur as exponents, then the answer is negative: *There is no general method to determine if an arbitrary exponential diophantine equation has a solution* (Section 11).

These decision problems are closely related to certain diophantine problems which are turned around from the usual problems. Ordinarily, a diophantine equation with, for example, one parameter a is given and we seek a simple characterization of the values of a for which the equation has a solution. Turning the problem around, we take some interesting set $\mathcal{S}$ of natural numbers† and seek a diophantine equation with one parameter a which has a solution for exactly those values of a which belong to $\mathcal{S}$. If there is such an equation, then we say that $\mathcal{S}$ is *diophantine.*

It turns out that *if the set of primes is diophantine, then there is no general method to determine if an arbitrary diophantine equation has a solution* (Section 12). Now it may seem unlikely that there is a diophantine equation with one parameter a which is solvable if and only if a is a prime. However, *there is an exponential diophantine equation with one parameter a which has a solution if and only if a is prime* (Section 10). Actually, much more is proved in Section 11, namely, *every recursively enumerable set is exponential diophantine.*

But *if the set of primes is not diophantine, then for each diophantine equation* $P(x, y, \cdots) = 0$, *either for every n there is a solution with $y \geqq x * n$ or there is some m such that every solution of $P = 0$ with $x > 1$ satisfies $y > x^m$* (Section 12). Here $x * n$ is the nth superpower of x, i.e., $*$ is the binary operation defined recursively by the equations $x * 0 = 1$ and $x * (n + 1) = x^{(x*n)}$.

Historical Remark: *Gödel's revolutionary work of* 1931 *permeates almost all of this study. The theory of computability was developed by Kleene, Turing, Post, and Church. Since this account is both an amalgamation and a specialization of their ideas, it is impossible to identify their contributions specifically and completely. The applications to diophantine equations have been made by Davis, Putnam, and myself.*

† The natural numbers are 0, 1, 2, $\cdots$.

2. PAIRS AND SEQUENCES

Cantor proved that the set of ordered pairs of natural numbers can be put in one-to-one correspondence with the set of natural numbers. This can be done in many ways, of course, but there is one particularly simple mapping function which we shall find useful. Let $J(x, y) = \frac{1}{2}((x + y)^2 + 3x + y)$. Then $J(x, y)$ is an integer-valued function which assumes each natural number as value exactly once for x and y natural numbers. In fact, $J(0, 0) = 0$ and

$$J(x, y) + 1 = \begin{cases} J(x + 1, y - 1), & \text{if } y > 0, \\ J(0, x + 1), & \text{if } y = 0. \end{cases}$$

So J maps the set of ordered pairs of natural numbers onto the natural numbers according to the following diagram.

	$J(x, y)$	y = 0	1	2	3	4
	0	0	1	3	6	10
	1	2	4	7	11	
x	2	5	8	12		
	3	9	13			
	4	14				

Two inverse functions K and L are uniquely determined by the equation $u = J(Ku, Lu)$.† The first few values of K and L which clearly indicate the general pattern of their sequence of values are given below:

u	0	1	2	3	4	5	6	7	8	9	10	11	12	13	14	15
Ku	0	0	1	0	1	2	0	1	2	3	0	1	2	3	4	0
Lu	0	1	0	2	1	0	3	2	1	0	4	3	2	1	0	5

† We usually omit parentheses in the functional notation for functions of one variable. Thus $F(x)$ is written Fx and $F(G(x))$ is written FGx.

To give an example of the use of pairing functions, suppose $\mathcal{S}$ is the range of a function $F(x, y)$ of two variables. Then $\mathcal{S}$ is also the range of $G(u) = F(Ku, Lu)$, a function of one variable. We can also recover the function F from G, since $F(x, y) = G(J(x, y))$.

We can also let a set $\mathcal{S}$ of natural numbers represent a binary relation $\mathcal{R}$ by the equivalence: $\mathcal{R}(x, y)$ if and only if $J(x, y) \in \mathcal{S}$. Thus, a sequence of sets $\mathcal{S}_0, \mathcal{S}_1, \mathcal{S}_2, \cdots$ is represented by a single set $\mathcal{S}$ through the correspondence $x \in \mathcal{S}_n$ if and only if $J(n, x) \in \mathcal{S}$.

Another example of the usefulness of pairing functions is a simple system of numbering diophantine equations. Here a diophantine equation is an equation of the form $F(x_0, x_1, \cdots) = G(x_0, x_1, \cdots)$ where F and G are terms built up from $x_0, x_1, \cdots$ and natural numbers by addition and multiplication. A diophantine equation can also be put in the form $P(x_0, x_1, \cdots) = 0$, where P is a polynomial with integer coefficients.

First we number the terms $\tau_0, \tau_1, \tau_2, \cdots$ built up from the variables and natural numbers by repeated additions and multiplications as follows:

$$\begin{aligned} \tau_{4n} &= n, \\ \tau_{4n+1} &= x_n, \\ \tau_{4n+2} &= \tau_{Kn} + \tau_{Ln}, \\ \tau_{4n+3} &= \tau_{Kn} \cdot \tau_{Ln}. \end{aligned}$$

Then we number the equations with the nth equation being $\tau_{Kn} = \tau_{Ln}$. We could also write the nth equation as $\tau_{Kn} - \tau_{Ln} = 0$. Thus, the eighth equation is $0 + 0 = x_0$ and the number of $2 \cdot x_0^2 = 1 + x_1^2$ is 7,697,614,550.

Finally, we wish to give a method of representing finite sequences of natural numbers due to Gödel. Let Rem (x, y) be the least nonnegative remainder of x divided by y.

LEMMA (GÖDEL 1931): *For every finite sequence* $s_0, s_1, \cdots, s_k$ *of natural numbers, there are natural numbers* a *and* d *such that*

$$s_t = \text{Rem}(a, 1 + (t + 1)d) \quad \textit{for} \quad t = 0, \cdots, k. \tag{2.1}$$

Proof: Now (2.1) is equivalent to

$$\begin{aligned} &a \equiv s_t \pmod{1 + (t + 1)d}, \\ &0 \leqq s_t < 1 + (t + 1)d \quad \text{for} \quad t = 0, \cdots, k. \end{aligned} \tag{2.2}$$

By the Chinese remainder theorem, the congruences can be satisfied for some a if the moduli are relatively prime and the inequalities will be satisfied if d is sufficiently large. Take for d a sufficiently large multiple of $k!$ so that $d > s_t$ for all $t \leqq k$. If a prime p divides both $1 + (t + 1)d$ and $1 + (t' + 1)d$, then $p \mid (t - t')d$. If $t \neq t'$, then $0 < |t - t'| \leqq k$. Hence $p|k!$ and so $p \mid d$. But $p \nmid d$ since by hypothesis $p \mid 1 + (t + 1)d$. Therefore for this choice of d, the moduli in (2.2) are relatively prime and a satisfying (2.2) exists.

This lemma is at the root of much of what follows.

3. COMPUTING AND LISTING

In Section 1 we spoke of a "general method" to tell if an arbitrary diophantine equation has a solution. What do we have in mind? We mean a set of instructions (necessarily finite) which describe in a completely deterministic way, how to start from an arbitrary diophantine equation $P(x_1, \cdots, x_k) = 0$ and to obtain after a finite number of steps the correct answer to the question: Does $P = 0$ have a solution? At no step in the process should the instructions call for either ingenuity or chance. On the other hand, we do not demand practicality of the method or place any restrictions on time or space needed to carry out the process.

Suppose we find a method to tell if an arbitrary diophantine equation has a solution. We give a proof that our method works, i.e., for every diophantine equation, it yields the correct answer to the question: Does this equation have a solution? Then if the argument is sound, everyone will presumably admit that Hilbert's problem is solved. We do not need to agree ahead of time on the exact meaning of the term "general method." But if there is no general method and we wish to prove that there is none, then we need to be precise. Hence until the notion of computability was defined, no one could ask if such a method exists.

Let us number all diophantine equations in some systematic way such as in the last section. Thus given any n, we can write down the nth equation and given an equation, we can figure out its number (or at least one of its numbers if some equations occur more than once). Let $\mathcal{S}$ be the set of numbers of equations which have solutions. Then a method to tell whether or not an arbitrary natural number n belongs to $\mathcal{S}$ would also give us a method to tell whether or not an arbitrary diophantine equation has a solution.

We call a set of numbers *computable* if there is a method of deciding whether or not an arbitrary natural number n belongs to the set. A *method* consists of a finite set of instructions which for each natural number n specifies a calculation terminating in a "yes" or "no" answer to the question: Does n belong to the set being computed?

Clearly, this is not a mathematical definition of the class of computable sets but, rather, it is a description of the intuitive concept of a computable set. In Section 5 we will give a strictly mathematical characterization of the sets which are identified with this intuitive concept of computable set.

We call a set $\mathcal{L}$ of natural numbers *listable* if there is a method of making a list of the elements of $\mathcal{L}$. A *list* is either a finite or infinite sequence, with or without repetitions. For example, the set $\mathcal{S}$ of numbers of solvable diophantine equations is listable. To make the list, we simply try the possible values of the arguments in the equations in a systematic way. Each time we find a solution of some equation, we put its number on our list.

What is a method of listing a set of numbers? It is a set of instructions which provides for a completely deterministic calculation which may or may not terminate. From time to time during the calculation, a particular number is designated as the next number on the list and we place it next on the list.

We say a function $F(x_1, \cdots, x_k)$ defined for all natural numbers and assuming natural numbers as values is *computable* if there is a method to compute the value of $F(x_1, \cdots, x_k)$ for an arbitrary k-tuple of natural numbers. Here a method is a set of instructions such that if it is applied to any k-tuple of natural numbers $x_1, \cdots, x_k$ it will yield a completely deterministic calculation of the natural number which is $F(x_1, \cdots, x_k)$.

We will take the concept of *listable set* as our fundamental intuitive notion. The following observations show that the concepts of computable set and computable function can be defined in terms of listable sets.

I. A set $\mathcal{S}$ of natural numbers is computable if and only if $\mathcal{S}$ and its complement, $\bar{\mathcal{S}}$, are both listable.

If we have instructions for computing $\mathcal{S}$, we can modify them trivially so that "yes" and "no" answers are interchanged. Thus we obtain instructions for computing $\bar{\mathcal{S}}$. Clearly we can list every computable set so both $\mathcal{S}$ and $\bar{\mathcal{S}}$ are listable. On the other hand, suppose $\mathcal{S}$ and $\bar{\mathcal{S}}$ are both infinite. If we have instructions for listing both $\mathcal{S}$ and $\bar{\mathcal{S}}$, we can make up a single list by alternately putting the next element from the list of $\mathcal{S}$ and then the next element from the list of $\bar{\mathcal{S}}$. To tell if n belongs to $\mathcal{S}$, we need only observe whether it occurs first in an even or odd position on our new list. (We know it will eventually occur and always at positions with the same parity.) Finally, if either $\mathcal{S}$ or $\bar{\mathcal{S}}$ is finite, then both are evidently computable.

In view of I and the earlier remark that the set of numbers of solvable diophantine equations is listable, the decision problem for diophantine equations is reduced to determining if the set of numbers of unsolvable diophantine equations is listable.

II. The range of a computable function is listable and conversely, every nonempty listable set $\mathcal{S}$ is the range of a computable function F.

We can list the values of a computable function Fx by computing $F0, F1, F2, \cdots$, in turn. Also, we can agree on an ordering of k-tuples in case the given function has k arguments and then list the functional values in the corresponding order. On the other hand, a nonempty listable set $\mathcal{S}$ is either finite so we can choose an ultimately constant function for F, or $\mathcal{S}$ is infinite and we take Fn equal to the nth number on a list of $\mathcal{S}$. In either case, F is computable.

III. A function Fx is computable if and only if its graph, i.e., the set $\{J(x, Fx)\}$, is listable.

If F is computable, then certainly $J(x, Fx)$ is also and by II, its range is listable. Suppose the graph of some function G is listable. To calculate Gn, we simply list the elements of the graph of G until some number u is listed with $Ku = n$. Then $Gn = Lu$.

[*Note*: Let the graph of $F(x, y)$ be $\{J(J(x, y), F(x, y))\}$. Then, in a similar way, we could prove III for functions of two variables and in fact, for any number of variables.]

We now assume that there is some mathematical language $\mathfrak{L}$ sufficiently rich that instructions can be given in $\mathfrak{L}$ for listing any listable set. Clearly we can assume that $\mathfrak{L}$ has only a finite number of symbols. Furthermore, we assume that we can recognize suitable instructions in $\mathfrak{L}$ and that the prescribed calculation consists of a sequence of steps of finite length. (This last condition means that if we decide to carry out one more step of the calculation we can eventually complete it.) In Section 5, we will justify these assumptions.

IV. There is a listable set $\mathfrak{U}$ such that $\{\mathfrak{U}_0, \mathfrak{U}_1, \cdots\}$ is the class of listable sets. [Here $k \in \mathfrak{U}_n$ if and only if $J(n, k) \in \mathfrak{U}$.]

We number the suitable sets of instructions in $\mathfrak{L}$: $\mathfrak{I}_0, \mathfrak{I}_1, \cdots$. (For example, we could first order them by the total number of symbols occurring in each set of instructions and then use a dictionary ordering among those of the same length.) Let the corresponding sets be $\mathfrak{U}_0, \mathfrak{U}_1, \cdots$. We now carry out the following sequence of calculations: We do one step of each set of instructions according to the sequence $\mathfrak{I}_0, \mathfrak{I}_0, \mathfrak{I}_1, \mathfrak{I}_0, \mathfrak{I}_1, \mathfrak{I}_2, \cdots$. When the calculation corresponding to $\mathfrak{I}_m$ yields the result that n is the next number on the list of $\mathfrak{U}_m$, we place $J(m, n)$ on our list of $\mathfrak{U}$. Hence $\mathfrak{U}$ is a listable set with the required property.

V. There is a listable set which is not computable.

Let $\mathfrak{U}$ be a set satisfying IV. Let $\mathfrak{D}$ be the diagonal set defined by $\mathfrak{D} = \{n | n \in \mathfrak{U}_n\}$. Then $\mathfrak{D}$ is certainly listable, since we can list the elements of $\mathfrak{U}$ and each time we come to a pair with equal components on the list of $\mathfrak{U}$ we can put their common value on the list of $\mathfrak{D}$. Suppose $\mathfrak{D}$ is computable; then $\overline{\mathfrak{D}}$ is listable, so $\overline{\mathfrak{D}} = \mathfrak{U}_k$ for some k. Hence for all n, $n \in \mathfrak{U}_k$ if and only if $n \notin \mathfrak{U}_n$. But this is impossible for $n = k$. Hence $\mathfrak{D}$ is not computable.

4. RECURSIVE FUNCTIONS

We now turn to the problem of defining mathematically a class of number-theoretic functions which we can identify with computable functions. We consider only functions defined for all natural numbers and whose values are natural numbers.

How do we compute familiar functions such as $[x/y]$, Rem (x, y), $\pi(x)$, $x * y$, the partition function, $\nu(n)$, etc.? We start with certain operations or initial functions, say addition, multiplication, and exponentiation, and then give formulas for computing new functions in terms of old by substitution and recursion (that is, computing the value of a function in terms of earlier values of the same function.)

A function is called *primitive recursive* if it can be obtained from certain initial functions by repeated substitution and recursion.

The initial functions are: the zero function of n variables O_n with $O_n(x_1, \cdots, x_n) = 0$ for $n \geqq 0$; the successor function S with Sx equal to the next number after x; and for every n, k with $1 \leqq k \leqq n$, the identity function I_{nk} of n variables defined by

$$I_{nk}(x_1, \cdots, x_n) = x_k.$$

The function F of n variables is said to be obtained by *substitution* from a function A of m variables and m functions $B_1, \cdots, B_m$ of n variables, provided

$$F(x_1, \cdots, x_n) = A(B_1(x_1, \cdots, x_n), \cdots, B_m(x_1, \cdots, x_n)).$$

In particular, we can obtain by substitution the functions $SO_n(x_1, \cdots, x_n)$, $SSO_n(x_1, \cdots, x_n)$, $\cdots$, i.e. all constant functions. Using identity functions, we can introduce extra variables or change the order of variables so that arbitrary substitutions may be made. For example, if we want to define a function F satisfying

$$F(x, y, z) = B(z, x),$$

we can use our substitution scheme by putting

$$F(x, y, z) = B(I_{33}(x, y, z), I_{31}(x, y, z)).$$

A function F of $n + 1$ variables is said to be obtained by *recursion* from a function A of n variables and a function B of $n+2$ variables, provided

$$F(x_1, \cdots, x_n, y) = A(x_1, \cdots, x_n),$$
$$F(x_1, \cdots, x_n, Sy) = B(x_1, \cdots, x_n, y, F(x_1, \cdots, x_n, y)).$$

For example, the functions $x + y$, $x \cdot y$, and x^y are primitive recursive as we see from the following sequence of equations:

$$x + 0 = x, \quad x + Sy = S(x + y);$$
$$x \cdot 0 = 0, \quad x \cdot Sy = x \cdot y + x;$$
$$x^0 = 1, \quad x^{Sy} = x^y \cdot x.$$

Note that $\operatorname{sgn} x = 0^{0^x}$ can be obtained from u^v by substitution. We can also define the predecessor function P by recursion: $P0 = 0$ and $PSx = x$. Let $x \dot{-} y$ be the function given by

$$x \dot{-} y = \begin{cases} x - y, & \text{if } x - y \geqq 0, \\ 0, & \text{otherwise.} \end{cases}$$

Using the predecessor function, we can obtain $x \dot{-} y$ by recursion as follows:

$$x \dot{-} 0 = x \quad \text{and} \quad x \dot{-} Sy = P(x \dot{-} y).$$

We obtain $|x - y| = (x \dot{-} y) + (y \dot{-} x)$ by substitution. Then, $\operatorname{Rem}(x, y)$ is defined by recursion as follows:

$$\operatorname{Rem}(0, y) = 0,$$
$$\operatorname{Rem}(Sx, y) = (S \operatorname{Rem}(x, y)) \cdot \operatorname{sgn} |y - S \operatorname{Rem}(x, y)|.$$

Suppose $A(x_1, \cdots, x_k, y)$ is primitive recursive, then the functions B and C given by

$$B(x_1, \cdots, x_k, z) = \sum_{y \leqq z} A(x_1, \cdots, x_k, y),$$

and

$$C(x_1, \cdots, x_k, z) = \prod_{y \leqq z} A(x_1, \cdots, x_k, y)$$

are also primitive recursive. In fact, they are obtained by recursion as follows:

$$B(x_1, \cdots, x_k, 0) = A(x_1, \cdots, x_k, 0),$$
$$B(x_1, \cdots, x_k, Sz) = B(x_1, \cdots, x_k, z) + A(x_1, \cdots, x_k, Sz);$$

$$C(x_1, \cdots, x_k, 0) = A(x_1, \cdots, x_k, 0),$$

$$C(x_1, \cdots, x_k, Sz) = C(x_1, \cdots, x_k, z) \cdot A(x_1, \cdots, x_k, Sz).$$

We can now define the function $\nu(n)$, which is the number of divisors of n for $n > 0$ and $\nu(0) = 1$,

$$\nu(n) = \sum_{x \leqq n} 0^{\text{Rem}(n,x)}.$$

Then

$$\pi(x) = \sum_{y \leqq x} 0^{|\nu(y)-2|}.$$

Actually all the functions mentioned in the second paragraph of this section can be shown to be primitive recursive. Indeed, all the functions which arise naturally in number theory turn out to be primitive recursive. It is only by deliberately making up a counter-example or defining a function metamathematically that we obtain a function which is not primitive recursive. Evidently, every primitive recursive function is computable since the definition of a function F by substitution or recursion provides a way to compute F.

Suppose we number all primitive recursive functions of one variable $F_0, F_1, \cdots$ systematically according to the way they are generated by substitution and recursion. Let G be the function defined by

$$G(t) = F_t(t) + 1.$$

Then G is not F_n for any n and hence G is not primitive recursive. But G is clearly computable.

A function F of n variables is *general recursive* if there are primitive recursive functions A of one variable and B of $n + 1$ variables such that

$$F(x_1, \cdots, x_n) = A\mu y\{B(x_1, \cdots, x_n, y) = 0\},$$

where for every n-tuple of natural numbers $x_1, \cdots, x_n$ there is some natural number y such that $B(x_1, \cdots, x_n, y) = 0$. Here $\mu y\{\cdots\}$ is the least number operator, i.e., $\mu y\{\cdots\}$ is the least number y which satisfies the condition in braces if there is such a y; otherwise $\mu y\{\cdots\}$ is undefined. This characterization of general recursive functions is due to Kleene and is usually called the Kleene normal form for general recursive functions.

Clearly, a general recursive function is computable. Namely, we compute $B(x_1, \cdots, x_n, 0), B(x_1, \cdots, x_n, 1), \cdots, B(x_1, \cdots, x_n, y)$ where $B(x_1, \cdots, x_n, y)$ is the first term equal to zero in this sequence. Since B is primitive recursive, we can compute each of these values. Then $F(x_1, \cdots, x_n) = Ay$, where A is also computable.

We cannot apply the diagonal argument to general recursive functions to show that some computable function is not general recursive because there is no method of listing all general recursive functions. More specifically, there is no method of telling which primitive recursive functions $B(x_1, \cdots, x_n, y)$ are such that for every n-tuple $x_1, \cdots, x_n$ of natural numbers there is a y with $B(x_1, \cdots, x_n, y) = 0$.

5. RECURSIVELY ENUMERABLE SETS

A nonempty set of natural numbers is *recursively enumerable* if it is the range of a primitive recursive function of one variable. The empty set is also taken to be recursive enumerable. A binary relation $\mathcal{R}$ is recursively enumerable if it is empty or if there are two primitive recursive functions F and G such that $\mathcal{R}(x, y)$ if and only if there is a natural number n such that $x = Fn$ and $y = Gn$.† Recursively enumerable relations of higher degree are defined similarly.

Every recursively enumerable set is listable by observation II of Section 3, since a primitive recursive function is computable. Conversely, we have the

CONTENTION: *Every listable set is recursively enumerable.*‡

† Equivalently, we could say $\mathcal{R}$ is recursively enumerable if the set $\{J(x, y) \mid \mathcal{R}(x, y)\}$ is recursively enumerable.

‡ This contention is equivalent to Church's thesis, i.e., every computable function is recursive.

We cannot prove this, since "listable" is an intuitive concept and not precisely defined. Why do we believe that every listable set is recursively enumerable? Not because computing successive values of a primitive recursive function is the most general method of making a list of numbers that we can think of but because of a metamathematical argument which can only be fully appreciated after some study of primitive recursive functions. Roughly, it goes like this. Any calculation can be broken down into a sequence of relatively simple basic steps. We can keep a record of the calculation so that at the end of each step, all the information needed to carry on an additional step has been recorded. Finally, if a step yields a number to be added to the list, we put the number at the end of the record of that step; if not, we put a period or other special symbol at the end. We assign numbers to the symbols being used including a number for each natural number. Next we assign numbers to all configurations of symbols in some simple systematic way according to their appearance. We can do this so that the function A given by

$$Ax = \begin{cases} n+1, & \text{if } x \text{ is the number of a configuration} \\ & \text{which ends in the natural number } n, \\ 0, & \text{otherwise,} \end{cases}$$

is primitive recursive. Let Bx be the number of the record of the calculation after x steps. (The record at any stage is a configuration of symbols.) Now $B(x + 1)$ is completely determined by the configuration whose number is Bx. Also $B(x + 1)$ is the number of the configuration obtained by making the small changes in the record due to one additional basic step of the calculation. Thus $B(x + 1) = CBx$ for some function C. Recalling how wide the class of primitive recursive functions is, we grant that C (and hence B) is primitive recursive. Let $\mathcal{A}$ be the set being listed and a some element of $\mathcal{A}$. Then $\mathcal{A}$ is the range of the primitive recursive function F given by

$$Fu = PABu + a \cdot 0^{ABu}.$$

Hence $\mathcal{A}$ is recursively enumerable.

It is easy to see directly that the range of a general recursive function is recursively enumerable. Suppose $Fx = A\mu y\{B(x, y) = 0\}$ where A and B are primitive recursive functions. Let

$$C(x, t) = \text{sgn} \prod_{z \leqq t} B(x, z).$$

Then, as we have seen, C is also primitive recursive. Let $k = \mu y\{B(0, y) = 0\}$. Then the range of F is also the range of the primitive recursive function G with

$$G(x, y) = A(0^{C(x,y)} \sum_{t \leqq y} C(x, t) + C(x, y) \cdot k).$$

Hence the range of F is the same as the range of $G(Ku, Lu)$, a primitive recursive function of one variable.

Actually, a much smaller class of functions than the class of primitive recursive functions is sufficient to obtain all recursively enumerable sets as ranges of functions of the class.

If the graph of a function F is recursively enumerable — say the range of A — then

$$Fx = LA\mu t\{KAt = x\}.$$

Hence F is general recursive. Conversely, the graph of a general recursive function F is the range of the general recursive function $J(x, Fx)$ and hence is recursively enumerable.

A set is *recursive* if both it and its complement are recursively enumerable.† In view of observations I–III of Section 3, recursive sets should be identified with computable sets and general recursive functions with computable functions.

There have been many different proposals for defining the class of listable sets.‡ It has turned out that the class of sets proposed has always been the same. Furthermore, the proofs that the proposed class is contained in the class of recursively enumerable sets can always be carried out according to the above sketch. Once the techniques of recursive function theory are familiar, these proofs are routine. And it is just this "routineness" which con-

† Equivalently, a set is recursive if its characteristic function is general recursive.

‡ A good account of several of these definitions is contained in Mendelson [10].

vinces us that the intuitive notion of "listable set" is correctly identified with recursively enumerable set.

6. GÖDEL'S THEOREM

We introduce logical symbols here as a necessary convenience although we do not need any knowledge of formal logic. We will use the universal quantifier $\bigwedge$ (for every), the existential quantifier $\bigvee$ (there exists), and the connectives $\wedge$ (and), $\vee$ (or), and $\neg$ (not). A formula is *arithmetical* if it is built up by means of logical symbols from equations of the form $a = b$, $a + b = c$, and $a \cdot b = c$ where a, b, and c are variables or symbols for particular numbers.

Examples The formula

$$(\bigwedge x)(\bigvee u, v, w, z)[x = u^2 + v^2 + w^2 + z^2]$$

expresses Lagrange's theorem that every natural number is the sum of four squares. The condition "p is a prime" is expressed by the formula

$$\neg (p^2 = p) \wedge \neg(\bigvee u, v)[p = (u + 2)(v + 2)].$$

Finally, the condition $y = [\sqrt{x}]$ is equivalent to the formula

$$(\bigvee u, v)[y^2 + u = x \wedge x + 1 + v = (y + 1)^2].$$

Relations defined by arithmetical formulas are called *arithmetical.* Gödel in 1931 proved the remarkable theorem that the relation given by an equation $F(x_1, \cdots, x_k) = y$ where F is any primitive recursive function is arithmetical. Consequently, every recursively enumerable set is arithmetical. Since as we have seen, there are some recursively enumerable sets which are not computable, there is some arithmetical formula Φ which defines a noncomputable set. Hence there is no decision method to answer questions of the sort: Does n satisfy Φ? Thus, there is no decision method to determine whether or not an arbitrary arithmetical statement is true.

In writing out arithmetical formulas, we can also use bounded quantifiers, i.e., $(\bigvee x \leqq y)[\cdots]$ (there exists an x not exceeding y such that $\cdots$) and $(\bigwedge x \leqq y)[\cdots]$ (for all x not exceeding y, $\cdots$). In fact, we could replace $(\bigvee x \leqq y)[\cdots]$ by $(\bigvee x, t)[x + t = y \wedge \cdots]$ and $(\bigwedge x \leqq y)[\cdots]$ by

$$(\bigvee x)(\bigvee t)[x = y + 1 + t \vee \cdots].$$

We shall now prove a strengthened form of Gödel's theorem† due to Martin Davis—the proof is essentially that of Gödel. In the next section, we will prove a still stronger theorem of Davis.

GÖDEL-DAVIS THEOREM: *Every relation given by* $F(x_1, \cdots, x_n) = y$, *where* F *is a primitive recursive function, can be expressed by an arithmetical formula in which the universal quantifiers are all bounded and there are no negation signs.*

For example, x is a power of 2 if and only if

$$(\bigwedge u \leqq x)(\bigwedge v \leqq x)[x \neq (2u + 3)v].$$

This formula can be put in the required form by eliminating the inequation. Namely, x is a power of 2 if and only if

$$(\bigwedge u \leqq x)(\bigwedge v \leqq x)(\bigvee w)[(x - (2u + 3)v)^2 = 1 + w].$$

Proof: We prove the theorem for functions F in the order in which they are constructed according to the definition of primitive recursive functions. That is, we check that the theorem holds if F is an initial function; then we show that the theorem holds for F if F is obtained by substitution or recursion from functions that satisfy the theorem.

I. If F is an initial function, then the theorem holds for F. In fact,

$$\begin{aligned} O_n(x_1, \cdots, x_n) = y &\quad\leftrightarrow\quad y = 0, \\ Sx = y &\quad\leftrightarrow\quad y = x + 1, \\ I_{nk}(x_1, \cdots, x_n) = y &\quad\leftrightarrow\quad y = x_k. \end{aligned}$$

Here we use $\leftrightarrow$ for "if and only if".

---

† This theorem is used to prove Davis' theorem (in the next section), which is needed only in Section 10. A reader unfamiliar with formal logic might do well to study the statement of Davis' theorem for the present and return to the proof later.

II. If $F(x_1, \cdots, x_n) = B(A_1(x_1, \cdots, x_n), \cdots, A_m(x_1, \cdots, x_n))$ where $A_1, \cdots, A_m$, and B satisfy the theorem, then F satisfies the theorem.

We have

$$F(x_1, \cdots, x_n) = y \leftrightarrow (\bigvee z_1, \cdots, z_m)[z_1 = A_1(x_1, \cdots, x_n) \wedge \cdots \wedge z_m = A_m(x_1, \cdots, x_n) \wedge y = B(z_1, \cdots, z_m)].$$

By the inductive hypothesis, each of the equations on the right side of the equivalence can be replaced by formulas of the required sort so F satisfies the theorem.

III. If $F(x_1, \cdots, x_k, 0) = A(x_1, \cdots, x_k)$ and $F(x_1, \cdots, x_k, y+1) = B(x_1, \cdots, x_k, y, F(x_1, \cdots, x_k, y))$, where A and B satisfy the theorem, then F satisfies the theorem.

Notice that for a fixed y, the sequence $u_0, \cdots, u_{y+1}$ is completely determined by the equations:

$$\begin{aligned} u_0 &= A(x_1, \cdots, x_k), \\ u_1 &= B(x_1, \cdots, x_k, 0, u_0), \\ u_2 &= B(x_1, \cdots, x_k, 1, u_1), \\ &\vdots \\ u_y &= B(x_1, \cdots, x_k, y-1, u_{y-1}), \\ u_{y+1} &= B(x_1, \cdots, x_k, y, u_y). \end{aligned} \tag{6.1}$$

(For convenience, we use one more equation than we need.) Hence if $u_0, \cdots, u_y, u_{y+1}$ satisfy (6.1), then $F(x_1, \cdots, x_k, t) = u_t$ for $t \leqq y + 1$. Now represent this sequence according to Gödel's lemma of Section 2. Thus, for some a and d

$$u_t = \text{Rem}(a, 1 + (t+1)d) \quad \text{for} \quad 0 \leqq t \leqq y + 1.$$

Finally,

$$\begin{aligned} &F(x_1, \cdots, x_k, y) = z \leftrightarrow \\ &\quad (\bigvee a, d)\{\text{Rem}\,(a, 1 + d) = A(x_1, \cdots, x_k) \\ &\quad \wedge z = \text{Rem}\,(a, 1 + (y+1)d) \\ &\quad \wedge (\bigwedge t \leqq y)[\text{Rem}\,(a, 1 + (t+2)d) \\ &\quad = B(x_1, \cdots, x_k, t), \text{Rem}\,(a, 1 + (t+1)d)]\}. \end{aligned} \tag{6.2}$$

Now

$$(6.3)\quad \text{Rem}\,(u, 1 + v) = w \leftrightarrow (\bigvee r, q)[u = (1 + v)r + w \wedge w + q = v].$$

Hence each of the equations in (6.2) can be replaced by a formula of the required sort, so F satisfies the theorem.

As an example of the proof of the theorem, we can write down the following definition of $y = 2^x$:

$$\begin{aligned} y = 2^x \leftrightarrow \\ &(\bigvee a, d)(\bigwedge t \leqq x)[\text{Rem}\,(a, 1 + d) = 1 \\ &\wedge \text{Rem}\,(a, 1 + (t + 2)d) = 2 \cdot \text{Rem}\,(a, 1 + (t + 1)d) \\ &\wedge y = \text{Rem}\,(a, 1 + (x + 1)d)]. \end{aligned}$$

Using (6.3), we can easily put the right side in the form required by the theorem.

7. DAVIS NORMAL FORM

In 1950, Davis considerably sharpened Gödel's result in the following theorem:

Every recursively enumerable relation can be put in the form

$$(7.1)\quad (\bigvee y)(\bigwedge u \leqq y)(\bigvee v_1 \leqq y) \cdots (\bigvee v_k \leqq y) [P(x_1, \cdots, x_n, y, u, v_1, \cdots, v_k) = 0],$$

where P is a polynomial with integer coefficients. Conversely, a relation defined by a formula of the form (7.1) is recursively enumerable.

We shall call this the Davis normal form of a recursively enumerable relation.

Proof: We start with the theorem of the last section. So we may assume the given recursively enumerable relation is given by a formula built up from diophantine equations by conjunctions, disjunctions, existential quantifiers, and bounded universal quantifiers. Now we show how various reductions can be made which lead to the Davis normal form.

1. If all the quantified variables are distinct, a formula of the Gödel type is equivalent to the formula obtained by putting all

the quantifiers at the beginning of the formula while maintaining their relative order and leaving the rest of the formula unchanged.

2. A formula built from equations by using conjunctions and disjunctions is equivalent to a polynomial equal to zero.

Since

$$A = 0 \wedge B = 0 \leftrightarrow A^2 + B^2 = 0$$

and

$$A = 0 \vee B = 0 \leftrightarrow A \cdot B = 0,$$

we can reduce the given formula step-wise to the form $P = 0$.

3. The relations $x > y$ and $x \geqq u$ are expressible as follows:

$$x > y \leftrightarrow (\bigvee z)[x = y + 1 + z]$$

and

$$x \geqq y \leftrightarrow (\bigvee z)[x = y + z].$$

Also we could just as well use the bounded quantifier $(\bigvee z \leqq x)$ in both formulas.

4. A formula of the form

$$(\bigwedge t \leqq x)(\bigvee u_1, \cdots, u_k)(\bigwedge z \leqq x')(\bigvee v_1, \cdots, v_l)[P = 0]$$

is equivalent to a formula in which the two universal quantifiers are adjacent:

$$\begin{gathered}(\bigvee a_1, \cdots, a_k, d_1, \cdots, d_k)(\bigwedge t \leqq x)(\bigwedge z \leqq x')(\bigvee u_1, \cdots, u_k, v_1, \cdots, v_l)\\ [u_1 = \text{Rem}(a_1, 1 + (t+1)d_1) \wedge \cdots\\ \wedge\, u_k = \text{Rem}\,(a_k, 1 + (t+1)d_k) \wedge P = 0].\end{gathered}$$

Note that $a_1, \cdots, a_k, d_1, \cdots, d_k$ can be chosen so that $u_1, \cdots, u_k$ are arbitrary numbers depending on t for each t with $t \leqq x$. Actually, we could take $d_1 = d_2 = \cdots = d_k$ since the conditions on d in Gödel's lemma (Section 2) could be satisfied by a single number. Using the equivalence

$$\text{Rem}\,(u, 1 + v) = w \leftrightarrow (\bigvee r, q)[u = (1 + v)r + w \wedge w + q = v],$$

and reductions 1 and 2, we can reduce the formula to one of the form

$$(\bigvee \cdots)(\bigwedge t \leqq x)(\bigwedge z \leqq x')(\bigvee \cdots)[P' = 0]$$

where P' is some polynomial with integer coefficients.

5. A formula of the form

$$(\bigwedge u \leqq x)(\bigwedge v \leqq x')(\bigvee w)[P = 0]$$

is equivalent to a formula with just one universal quantifier, namely,

$$(\bigvee z)(\bigwedge r \leqq z)(\bigvee u, v, w) \\ [z = J(x, x') \wedge r = J(u, v) \wedge (u > x \vee v > x' \vee P = 0)].$$

Here we use the fact that if $u \leqq x$ and $v \leqq x'$, then $J(u, v) \leqq J(x, x')$. Also $z = J(x, x')$ is equivalent to the equation $2z = 2J(x, x')$ with integral coefficients. Using reductions 2 and 3, we obtain an equivalent formula of the form

$$(\bigvee z)(\bigwedge r \leqq z)(\bigvee u, v, w, s, t)[P = 0],$$

where P is some new polynomial.

By repeated use of reductions 4 and 5, we can obtain a formula in the form

$$(\bigvee x_1, \cdots, x_k)(\bigwedge u \leqq z)(\bigvee v_1, \cdots, v_l)[P = 0].$$

6. A formula of the form

$$(\bigvee x, y)(\bigwedge u \leqq z)(\bigvee v_1, \cdots, v_k)[P = 0]$$

is equivalent to a formula with just one initial existential quantifier:

$$(\bigvee t)(\bigwedge u \leqq t)(\bigvee v_1, \cdots, v_k, x, y) \\ [t = J(J(x, y), z) \wedge (u > z \vee P = 0)].$$

Now by repeated applications of reduction 6, we will obtain a formula of the form

$$(\bigvee w)(\bigwedge u \leqq w)(\bigvee v_1, \cdots, v_r)[P = 0].$$

This formula is equivalent to

$$(\bigvee y)(\bigwedge u \leqq y)(\bigvee v_1 \leqq y) \cdots (\bigvee v_r \leqq y)(\bigvee w \leqq y)(\bigvee z \leqq y)(\bigvee t \leqq y) \\ [y = J(w, z) \wedge (u = w + 1 + t \vee P = 0)].$$

Finally, by application of reduction 2 to the last formula, we obtain the Davis normal form.

Conversely, if $\mathcal{S}$ is defined by the equivalence

$$x \in \mathcal{S} \leftrightarrow (\bigvee y)(\bigwedge u \leqq y)(\bigvee v_1 \leqq y) \cdots (\bigvee v_k \leqq y)[P = 0],$$

then $\mathcal{S}$ is recursively enumerable. Let $G(x, y)$ be defined by

$$G(x, y) = \sum_{u \leqq y} \prod_{v_1 \leqq y} \cdots \prod_{v_k \leqq y} P^2(x, y, u, v_1, \cdots, v_k);$$

then G is primitive recursive and

$$x \in \mathcal{S} \leftrightarrow (\bigvee y)[G(x, y) = 0].$$

Hence if $\mathcal{S}$ is not empty, let a be some particular element of $\mathcal{S}$. Then $H(x, y)$ defined by

$$H(x, y) = 0^{G(x,y)} \cdot x + \operatorname{sgn} G(x, y) \cdot a$$

is primitive recursive. Clearly, $\mathcal{S}$ is the range of H, so $\mathcal{S}$ is recursively enumerable.

8. DIOPHANTINE RELATIONS

By a solution of a diophantine equation, we mean a solution in natural numbers. Ordinarily, integer solutions are sought. However the answer to the decision problem is the same in the two cases and it is more convenient for us to consider only natural number solutions. Indeed, $P(x_1, \cdots, x_k) = 0$ has a solution in integers if and only if one of the equations $P(\pm x_1, \cdots, \pm x_k) = 0$ given by a particular choice of signs has a solution in natural numbers. Hence if Q is the polynomial obtained by multiplying together all the 2^k polynomials $P(\pm x_1, \cdots, \pm x_k)$ obtained by different choices of signs, then $P(x_1, \cdots, x_k) = 0$ has a solution in integers if and only if $Q(x_1, \cdots, x_k) = 0$ has a solution in natural numbers. On the other hand, in view of Lagrange's theorem, $P(x_1, \cdots, x_k) = 0$ has a solution in natural numbers if and only if $P(u_1^2 + v_1^2 + w_1^2 + z_1^2, \cdots, u_k^2 + v_k^2 + w_k^2 + z_k^2) = 0$ has a solution in integers.

A set $\mathcal{S}$ of natural numbers is *diophantine* if there is a polynomial $P(n, x_1, \cdots, x_k)$ with integer coefficients such that n belongs to $\mathcal{S}$ if and only if there are natural numbers $x_1, \cdots, x_k$ with $P(n, x_1, \cdots, x_k) = 0$. Similarly, a binary relation $\mathcal{R}(m, n)$ is *diophantine* if and only if there is a polynomial $P(m, n, x_1, \cdots, x_k)$ with integer coefficients such that $\mathcal{R}(m, n)$ if and only if there are natural numbers $x_1, \cdots, x_k$ such that $P(m, n, x_1, \cdots, x_k) = 0$, and so on for relations of higher degree.

Examples. The set of perfect squares is diophantine, since n is a square if and only if $n - x^2 = 0$ has a solution in natural numbers. The set of numbers which are not squares is also diophantine. For n is not a square if and only if it lies strictly between two consecutive squares. Thus n is not a square if and only if the system of equation

$$\begin{aligned} x^2 + 1 + y &= n, \\ n + 1 + z &= (x + 1)^2, \end{aligned}$$

has a solution for x, y, z in natural numbers. Furthermore, replacing the system of equations by a single equation, we have n is not a square if and only if the equation

$$(x^2 + 1 + y - n)^2 + (n + 1 + z - (x + 1)^2)^2 = 0$$

has a solution for x, y, z in natural numbers. A simpler equation with the same property can be obtained from Pell's equation. Namely, n is not a square if and only if the equation $(x + 2)^2 - ny^2 = 1$ has a solution for x and y in natural numbers.

The relations $m \mid n$ and $m \nmid n$ are both diophantine. Clearly, $m \mid n$ if and only if the equation $mx = n$ has a solution for x in natural numbers. Also $m \nmid n$ if and only if there are natural numbers x and y such that either $n = mx + y$ and $0 < y < m$, or $m = 0$ and $n > 0$. Hence $m \nmid n$ if and only if the equation

$$((n - mx - y)^2 + (y - 1 - z)^2 + (m - y - 1 - u)^2) \\ (m^2 + (n - 1 - v)^2) = 0$$

has a solution for x, y, z, u, and v in natural numbers.

Finally, the relation $(m, n) = 1$ holds if and only if there are natural numbers u and v with $mu = 1 + nv$, so the relation of relative primeness is diophantine.

Another characterization of diophantine sets was found by Putnam [12].† Suppose $\mathcal{S}$ is the set of natural numbers n such that $P(n, x_1, \cdots, x_k) = 0$ has a solution for $x_1, \cdots, x_k$ in natural numbers. Let

† Putnam also shows that there is no method of telling whether or not an arbitrary diophantine equation with one parameter a is solvable for every value of a.

$$P_0(x_0, x_1, \cdots, x_k) = (x_0 + 1)(1 - P^2(x_0, x_1, \cdots, x_k)) - 1$$

Then $P_0(n, x_1, \cdots, x_k) \geqq 0$ if and only if $P(n, x_1, \cdots, x_k) = 0$. Furthermore, if $P(n, x_1, \cdots, x_k) = 0$ then $P_0(n, x_1, \cdots, x_k) = n$. Hence a natural number n belongs to $\mathcal{S}$ if and only if n belongs to the range of P_0.

9. EXPONENTIAL DIOPHANTINE RELATIONS

We have defined a diophantine equation to be an equation which can be put in the form $A = B$ where A and B are terms built from natural numbers and variables by addition and multiplication. Similarly, we define an *exponential diophantine equation* to be an equation which can be put in the form $E = F$ where E and F are terms built from natural numbers and variables by addition, multiplication, and exponentiation.

A set $\mathcal{S}$ of natural numbers is *exponential diophantine* if there is an exponential diophantine equation $E(n, x_1, \cdots, x_k) = F(n, x_1, \cdots, x_k)$ which has a solution for $x_1, \cdots, x_k$ in natural numbers if and only if n belongs to $\mathcal{S}$. Exponential diophantine relations are defined similarly.

Examples

1. The relation $m = \binom{n}{r}$ is exponential diophantine. In fact, we shall show that

$$(9.1) \quad \binom{n}{r} = [2^{nr}(1 + 2^{-n})^n] - 2^n[2^{n(r-1)}(1 + 2^{-n})^n] + 0^{n+r}.$$

Suppose $n > 0$ and $r > 0$. Then

$$2^{nr}(1 + 2^{-n})^n = 2^{nr} \sum_{t=0}^{n} \binom{n}{t} 2^{-nt} < \sum_{t=0}^{r} \binom{n}{t} 2^{n(r-t)} + \frac{(2^n - 1)}{2^n}.$$

Hence

$$[2^{nr}(1 + 2^{-n})^n] = \sum_{t=0}^{r} \binom{n}{t} 2^{n(r-t)}.$$

Similarly,

$$2^n[2^{n(r-1)}(1 + 2^{-n})^n] = \sum_{t=0}^{r-1} \binom{n}{t} 2^{n(r-t)}.$$

So (9.1) holds for $n, r > 0$. We can easily check that it also holds when either $n = 0$ or $r = 0$.

To express the relation $m = \binom{n}{r}$ by an exponential diophantine equation we use the fact that for $b > 0$, $[a/b] = c$ if and only if $bc \leqq a < b(c + 1)$. Hence $m = \binom{n}{r}$ if and only if the following conditions can be satisfied by natural numbers x and y:

$$\begin{aligned} m &= x - 2^n y + 0^{n+r}, \\ 2^{n^2}x &\leqq 2^{nr}(2^n + 1)^n < 2^{n^2}(x + 1), \\ 2^n 2^{n^2} y &\leqq 2^{nr}(2^n + 1)^n < 2^n 2^{n^2}(y + 1). \end{aligned}$$

Thus, $m = \binom{n}{r}$ if and only if the following system of equations has a solution for x, y, u, v, w, and z in natural numbers:

$$\begin{aligned} m + 2^n y &= x + 0^{n+r}, \\ 2^{n^2}x + u &= 2^{nr}(2^n + 1)^n, \\ 2^{nr}(2^n + 1)^n + 1 + v &= 2^{n^2}(x + 1), \\ 2^n 2^{n^2} y + w &= 2^{nr}(2^n + 1)^n, \\ 2^{nr}(2^n + 1)^n + 1 + z &= 2^n 2^{n^2}(y + 1). \end{aligned}$$

Now let these five equations be $E_i = F_i$ $(i = 1, \cdots, 5)$. Let

$$G = \sum_{i=1}^{5} (E_i - F_i)^2.$$

Then $G = 0$ is an exponential diophantine equation with three parameters m, n, and r which has a solution in natural numbers if and only if $m = \binom{n}{r}$.

2. The relation $m = n!$ is exponential diophantine. To see this, we use the fact that for s sufficiently large compared to n,

$$n! = \left[\frac{s^n}{\binom{s}{n}} \right]. \tag{9.2}$$

In fact, it is easy to check that (9.2) holds whenever $s > (2n)^{n+1}$. For $n > 0$,

$$n! \leqq \frac{s^n}{\binom{s}{n}} = n!\left(\frac{s}{s-1}\right)\left(\frac{s}{s-2}\right)\cdots\left(\frac{s}{s-n+1}\right),$$

$$< n!\left(\frac{1}{1-\frac{n}{s}}\right)^n.$$

We now use the elementary inequalities:

$$\frac{1}{1-\theta} < 1 + 2\theta \quad \text{for} \quad 0 < \theta < \frac{1}{2},$$

and

$$(1+\theta)^k < 1 + 2^k\theta \quad \text{for} \quad 0 < \theta < 1.$$

Thus,

$$\frac{s^n}{\binom{s}{n}} < n!\left(1 + \frac{2n}{s}\right)^n < n!\left(1 + 2^n\left(\frac{2n}{s}\right)\right)$$

$$= n! + \frac{2^n(2n)n!}{s}$$

$$< n! + 1.$$

Hence (9.2) is established for $s > (2n)^{n+1}$. Finally, $m = n!$ if and only if there are natural numbers r, s, and t such that

$$s > (2n)^{n+1},$$

$$t = \binom{s}{n},$$

$$s^n = mt + r,$$

$$r < t.$$

Since each of these conditions is exponential diophantine, the relation $m = n!$ is also exponential diophantine.

3. The set of primes is exponential diophantine.

Since p is prime if $p = x + 1$ and $p \nmid x!$ for some natural number x and both of these conditions are exponential diophantine, the set of primes is also. That is, there is an exponential diophantine equation with one parameter p which has a solution if and only if p is a prime. We could write out such an equation, as we have indicated, but it would be rather complicated. It would be interesting to find a simple equation which would define the set of primes.

4. (Davis and Putnam). The relation given by

$$m = \prod_{k=1}^{n} (a + bk)$$

is exponential diophantine.

Now for $b > 0$,

$$\prod_{k=1}^{n} (a + bk) = \binom{n + a/b}{n} b^n n!.$$

(The binomial coefficient $\binom{\alpha}{n} = \dfrac{\alpha(\alpha - 1)\cdots(\alpha - n + 1)}{n!}$ for rational α.) Hence it will be sufficient to show that the relation $R(s, t, p, q, r)$ which holds among the natural numbers s, t, p, q, and r if and only if

$$\frac{s}{t} = \binom{\frac{p}{q}}{r}, \quad (s, t) = 1, \quad \text{and} \quad \frac{p}{q} > r. \tag{9.3}$$

Let α be a rational number, r a natural number less than α, and x a real number with $0 < x < 1$. Then

$$(1 + x)^\alpha = \sum_{j=0}^{r} \binom{\alpha}{j} x^j + \binom{\alpha}{r + 1} x^{r+1}(1 + \theta x)^{\alpha - r - 1}$$

for some θ with $0 < \theta < 1$. Let $\alpha = p/q > r$ and $x = a^{-2}$ where a is an integer greater than 1. Then

$$a^{2r+1}(1 + a^{-2})^\alpha = \sum_{j=0}^{r} \binom{\alpha}{j} a^{2r-2j+1} + \binom{\alpha}{r + 1} a^{-1}(1 + \theta a^{-2})^{\alpha - r - 1}. \tag{9.4}$$

Let

$$S_r^\alpha(a) = \sum_{j=0}^{r} \binom{\alpha}{j} a^{2r-2j+1}. \tag{9.5}$$

Then we can define $\binom{\alpha}{r}$ explicitly in terms of these sums, since

$$(9.6) \qquad \binom{\alpha}{r} = a^{-1}S_r^\alpha(a) - aS_{r-1}^\alpha(a).$$

Furthermore, since the error term in (9.4) is nonnegative for $\alpha \geqq r$, (9.4) can be rewritten as follows:

$$(9.7) \qquad S_r^\alpha(a) = a^{2r+1}(1 + a^{-2})^\alpha - \theta'\alpha^{r+1}a^{-1}2^\alpha$$

for some θ' with $0 \leqq \theta' \leqq 1$. Now if $q^r r! | a$, then both $S_r^{p/q}(a)$ and $S_{r-1}^{p/q}(a)$ will be integers. Also if $a > p^{r+1}2^p$ then the error term in (9.7) is less than 1. Hence $\mathcal{R}(x, y, p, q, r)$ holds if and only if there is a natural number a so that

$$\frac{p}{q} > r,$$
$$(x, y) = 1,$$
$$a > p^{r+1}2^p,$$
$$q^r r! \mid a,$$
$$\frac{ax}{y} = [a^{2r+1}(1 + a^{-2})^{p/q}] - a^2[a^{2r-1}(1 + a^{-2})^{p/q}].$$

It is easy to see that each of these conditions is exponential diophantine,† so $\mathcal{R}$ and hence the relation $m = \prod_{k=1}^{n} (a + bk)$ are exponential diophantine. As an immediate consequence, we see that the relation $m = \prod_{k=0}^{n} (a - k)$ is exponential diophantine. Indeed, $m = \prod_{k=0}^{n} (a - k)$ if and only if there is an x such that

$$(a + x = n \wedge m = 0) \vee (a = n + x + 1 \wedge m = \prod_{k=1}^{n+1} (x + k)).$$

Since both of the alternatives are exponential diophantine, we can use the fact that $A = 0 \vee B = 0$ if and only if $AB = 0$ to see that $m = \prod_{k=0}^{n} (a - k)$ is exponential diophantine.

---

† Only the last condition requires simplification. We can use the fact that $u = [vw^{p/q}]$ if and only if $u^q \leqq v^q w^p$ and $v^q w^p < (u + 1)^q$.

10. EXPONENTIAL DIOPHANTINE = RECURSIVELY ENUMERABLE

It is immediately evident that every exponential diophantine relation is listable. It is enough to search systematically for solutions to the defining equation and record the successes. Hence every exponential diophantine relation is recursively enumerable. But since this is a mathematical statement, we should not rely on the contention of Section 5. Suppose n belongs to $\mathcal{S}$ if and only if $E(n, x_1, \cdots) = 0$ has a solution for $x_1, \cdots$ in natural numbers and $\mathcal{S}$ is not empty. Then $\mathcal{S}$ is the range of the primitive recursive function

$$0^{E(n,x_1,\cdots)^2} \cdot n + \operatorname{sgn}(E(n, x_1, \cdots)^2) \cdot s$$

where s is some particular element of $\mathcal{S}$.

THEOREM (DAVIS, PUTNAM, ROBINSON): *Every recursively enumerable relation is exponential diophantine.*

We shall first prove a lemma which contains the heart of the argument. Let $\mathfrak{x}$ stand for $x_1, \cdots, x_n$.

LEMMA: *Let* $P(\mathfrak{x}, y, u, v_1, \cdots, v_k)$ *be a polynomial with integer coefficients. Let* $Q(\mathfrak{x}, y)$ *be a polynomial such that*

$$Q(\mathfrak{x}, y) \geqq y \tag{10.1}$$

and

$$(\bigwedge u \leqq y)(\bigwedge v_1 \leqq y)\cdots(\bigwedge v_k \leqq y) \quad [|P(\mathfrak{x}, y, u, v_1, \cdots, v_k)| \leqq Q(\mathfrak{x}, y)]. \tag{10.2}$$

Then

$$(\bigwedge u \leqq y)(\bigvee v_1 \leqq y)\cdots(\bigvee v_k \leqq y) \quad [P(\mathfrak{x}, y, u, v_1, \cdots, v_k) = 0] \tag{10.3}$$

is equivalent to

$$\begin{aligned}(\bigvee c, t, a_1, \cdots, a_k)[t &= Q(\mathfrak{x}, y)! \\ \wedge\ 1 + (c+1)t &= \prod_{m \leqq y} (1 + (m+1)t) \\ \wedge\ 1 + (c+1)t &\mid P(\mathfrak{x}, y, c, a_1, \cdots, a_k) \\ \wedge\ 1 + (c+1)t &\mid \prod_{j \leqq y} (a_1 - j) \wedge \cdots \\ \wedge\ 1 + (c+1)t &\mid \prod_{j \leqq y} (a_k - j)].\end{aligned} \tag{10.4}$$

[Here k is a particular natural number depending only on the number of variables in P. Hence for each polynomial P, the formula (10.4) has a fixed number of terms.]

Proof of lemma: The conditions

$$(10.5)\quad t = Q(\mathfrak{x}, y)!, \quad 1 + (c+1)t = \prod_{m \leqq y} (1 + (m+1)t)$$

uniquely determine t and c. From (10.1), (10.2), and (10.5), we have for $m, m' \leqq y$ and $m \neq m'$,

$$(10.6)\qquad (1 + (m+1)t, 1 + (m'+1)t) = 1$$

and

$$(10.7)\qquad c \equiv m \pmod{1 + (m+1)t}.$$

Also if p is prime and $p \mid 1 + (m+1)t$, then $p > y$ and $p > |P(\mathfrak{x}, y, u, v_1, \cdots, v_k)|$ for all $u, v_1, \cdots, v_k$ which do not exceed y.

I. Suppose the condition in brackets in (10.4) holds. Let $u \leqq y$ and p be a prime divisor of $1 + (u+1)t$. Put $v_i = \text{Rem}\,(a_i, p)$ for $i = 1, \cdots, k$. Now since

$$1 + (c+1)t \mid \prod_{j \leqq y} (a_i - j),$$

there is some $j \leqq y$ so that $p \mid a_i - j$. Hence $v_i \leqq y$. From (10.7), $c \equiv u \pmod{p}$ and by definition, $a_i \equiv v_i \pmod{p}$. Hence

$$P(\mathfrak{x}, y, u, v_1, \cdots, v_k) \equiv P(\mathfrak{x}, y, c, a_1, \cdots, a_k) \pmod{p}.$$

But by (10.4),

$$P(\mathfrak{x}, y, c, a_1, \cdots, a_k) \equiv 0 \pmod{p}.$$

So $P(\mathfrak{x}, y, u, v_1, \cdots, v_k) \equiv 0 \pmod{p}$. We have already seen that

$$|P(\mathfrak{x}, y, u, v_1, \cdots, v_k)| < p,$$

so

$$P(\mathfrak{x}, y, u, v_1, \cdots, v_k) = 0.$$

II. Suppose (10.3) holds. Let c and t be determined by (10.5). Let $v_{ui} \leqq y$ be chosen for $u \leqq y$ and $i = 1, \cdots, k$ so that $P(\mathfrak{x}, y, u, v_{u1}, \cdots, v_{uk}) = 0$. Let $a_1, \cdots, a_k$ satisfy

$$a_i \equiv v_{ui} \pmod{1 + (u+1)t} \quad \text{for } u \leqq y.$$

There are such $a_1, \cdots, a_k$ by the Chinese remainder theorem since the moduli are relatively prime by (10.6). Hence

$$\begin{aligned} P(\mathfrak{x}, y, c, a_1, \cdots, a_k) &\equiv P(\mathfrak{x}, y, u, v_{u1}, \cdots, v_{uk}) \\ &\equiv 0 \pmod{1 + (u + 1)t}. \end{aligned}$$

But since the moduli are relatively prime, it follows that

$$\prod_{u \leqq y} (1 + (u + 1)t) \mid P(\mathfrak{x}, y, c, a_1, \cdots, a_k),$$

i.e., $1 + (c + 1)t \mid P(\mathfrak{x}, y, c, a_1, \cdots, a_k)$. Furthermore,

$$1 + (u + 1)t \mid a_i - v_{ui},$$

so

$$1 + (u + 1)t \mid a_i - j, \quad \text{for some } j \leqq y.$$

But since the divisors are relatively prime by (10.6), we have

$$1 + (c + 1)t \mid \prod_{j \leqq y} (a_i - j), \quad \text{for } i = 1, \cdots, k.$$

Hence (10.4) is satisfied and the lemma has been established.

Proof of Theorem: Let $\mathfrak{R}(x_1, \cdots, x_n)$ be a recursively enumerable relation. Then by Davis' result, there is a polynomial $P(\mathfrak{x}, y, u, v_1, \cdots, v_k)$ so that

$$\mathfrak{R}(x_1, \cdots, x_n)$$

holds if and only if

$$(\bigvee y)(\bigwedge u \leqq y)(\bigvee v_1 \leqq y) \cdots (\bigvee v_k \leqq y)[P(\mathfrak{x}, y, u, v_1, \cdots, v_k) = 0].$$

Let $Q(\mathfrak{x}, y)$ be the polynomial obtained from P by replacing the coefficients by their absolute value, substituting y for each of the variables $u, v_1, \cdots, v_k$, and adding y to the resulting polynomial. Clearly, Q satisfies the condition of lemma. Hence $\mathfrak{R}(x_1, \cdots, x_n)$ holds if and only if there is a natural number y satisfying the relation given by (10.4). Since each of the relations occurring in (10.4) is exponential diophantine, $\mathfrak{R}$ is exponential diophantine.

We return to the intuitive theory discussed in Sections 3 and 4 for an application of the theorem.

COROLLARY: *There is a particular exponential diophantine equation $E(n, x_1, \cdots, x_k) = 0$ for which there is no general method to tell for an arbitrary natural number n whether or not $E(n, x_1, \cdots, x_k) = 0$ has a solution for $x_1, \cdots, x_k$ in natural numbers.*

Proof: We have seen that there is a listable set $\mathcal{S}$ which is not computable. By contention, every listable set is recursively enumerable and hence exponential diophantine. Thus there is an exponential diophantine equation $E(n, x_1, \cdots, x_k) = 0$ such that n belongs to $\mathcal{S}$ if and only if $E = 0$ has a solution. Hence there can be no general method of telling, for an arbitrary n, whether $E(n, x_1, \cdots, x_k) = 0$ has a solution for $x_1, \cdots, x_k$ in natural numbers.

REMARKS: *This is a stronger result than the fact that there is no general method to tell whether or not an arbitrary exponential diophantine equation has a solution. In the case of diophantine equations, it may be that every diophantine set is computable but still there is no method to tell if an arbitrary diophantine equation has a solution.*

What relation does this result have to Gödel's incompleteness theorem? Given any effective system of true axioms, Gödel showed how to construct a particular arithmetical statement which is clearly true but not provable from the given axioms. By an arithmetical statement, we mean a statement that can be expressed by an arithmetical formula (Section 6).

Suppose we number exponential diophantine equations as we did ordinary diophantine equations in Section 2, say $E_0(x_0, x_1, \cdots) = 0$, $E_1(x_0, x_1, \cdots) = 0, \cdots$. Let $\mathcal{A}$ be some effectively given axiom system. We can assume either that the axioms are true or that the system is consistent and strong enough to prove the solvability of every exponential diophantine equation which has in fact a solution.† Let $\mathcal{U}$ be the set of natural numbers n such that $E_n(n, x_1, \cdots) = 0$ can be proved to have no solution for $x_1, \cdots$ in natural numbers on the basis of the axiom system $\mathcal{A}$. The set $\mathcal{U}$ is certainly listable, since we can systematically run through the possible proofs and put n on our list each time we conclude $E_n(n, x_1, \cdots) = 0$ is unsolvable. Hence there is some natural number k such that $n \in \mathcal{U}$ if and only if $E_k(n, x_1, \cdots) = 0$

† Note that the axioms need not be limited to number theory. They can include axioms of analysis, set theory, etc.

has a solution for $x_1, \cdots$. That is, the unsolvability of $E_n(n, x_1, \cdots) = 0$ is provable from $\mathfrak{A}$ if and only if $E_k(n, x_1, \cdots) = 0$ is solvable. Putting $n = k$, we see that the unsolvability of $E_k(k, x_1, \cdots) = 0$ is provable if and only if $E_k(k, x_1, \cdots) = 0$ is solvable. If the axioms are true and $E_k(k, x_1, \cdots) = 0$ is solvable, or if the axioms are consistent and $E_k(k, x_1, \cdots) = 0$ is solvable and provably so, then $E_k(k, x_1, \cdots) = 0$ cannot be proved unsolvable. Hence $E_k(k, x_1, \cdots) = 0$ is unsolvable but cannot be proved to be unsolvable.

Now given $\mathfrak{A}$ effectively, we can actually calculate k. Briefly, Gödel showed how to obtain a primitive recursive function whose range is $\mathfrak{U}$.† We can then use the reductions of Sections 6 and 7 to obtain $\mathfrak{U}$ in the Davis normal form. Finally, the theorem of this section yields the required exponential diophantine equation and we can calculate its number k. Actually to carry out this program of obtaining the required exponential diophantine equation for a nontrivial axiom system would hardly be worthwhile since the complexity of the resulting equation would outweigh its interest.

Thus we have sketched the proof that *to any effectively given system of true axioms, there is a method for finding an exponential diophantine equation which has no solution but for which there is no proof from the given axioms that the equation is unsolvable.*

11. REDUCTION OF EXPONENTIAL DIOPHANTINE EQUATIONS

Is the relation $r = s^t$ diophantine? If so, then every exponential diophantine equation could be replaced by an equivalent diophantine equation in more variables. Also every recursively enumerable relation would be diophantine and hence Hilbert's problem would be unsolvable. At present, we don't know the answer to this question. We shall show now that if any infinite set of primes is diophantine or if any relation of roughly expo-

† Essentially, his argument is the one sketched in Section 5 showing that any listable set is recursively enumerable.

nential order of growth is diophantine, then $r = s^t$ is indeed diophantine.

We need a few simple lemmas. In the following, let $a > 1$ and a_n, a'_n be defined by the equation

$$a_n + a'_n\sqrt{a^2 - 1} = (a + \sqrt{a^2 - 1})^n. \tag{11.1}$$

Thus,

$$\begin{aligned} a_0 &= 1 & a'_0 &= 0, \\ a_1 &= a, & a'_1 &= 1, \\ a_{n+2} &= 2a \cdot a_{n+1} - a_n, & a'_{n+2} &= 2a \cdot a'_{n+1} - a'_n. \end{aligned} \tag{11.2}$$

LEMMA 1: *Natural numbers x and y satisfy the equation $x^2 - (a^2 - 1)y^2 = 1$ if and only if there is a natural number n such that $x = a_n$ and $y = a'_n$.*

This is a special case of the familiar result concerning Pell's equation $x^2 - dy^2 = 1$, where d is not a perfect square. Namely, x and y satisfy $x^2 - dy^2 = 1$ if and only if for some n, $x + \sqrt{d}y = (x_1 + \sqrt{d}y_1)^n$, where x_1, y_1 is the least positive solution of the equation which always has a nontrivial solution if d is not a square (cf. LeVeque [9], p. 116.) In our case, $a > 1$ so that $a^2 - 1$ is not a square and $x = a$, $y = 1$ is clearly the least positive solution.

LEMMA 2: *For $a > 1$,*

$$a'_n \equiv n \pmod{a - 1} \tag{11.3}$$

and

$$a_n - a'_n(a - s) \equiv s^n \pmod{2as - s^2 - 1}. \tag{11.4}$$

Proof: The proof is by induction on n. From (11.2), we see that both (11.3) and (11.4) hold for $n = 0$ and $n = 1$. Assume (11.3) and (11.4) have been proved for n and $n + 1$, then

$$a'_{n+2} \equiv 2a \cdot (n + 1) - n \equiv n + 2 \pmod{a - 1}$$

and

$$\begin{aligned} a_{n+2} - a'_{n+2}(a - s) &\equiv 2a(a_{n+1} - a'_{n+1}(a - s)) - (a_n - a'_n(a - s)), \\ &\equiv 2as^{n+1} - s^n \equiv s^{n+2} \pmod{2as - s^2 - 1}. \end{aligned}$$

LEMMA 3. *There is a diophantine relation $\mathfrak{Q}(a, u)$ such that*

(i) *if* $\mathcal{Q}(a, u)$ *then* $u \geqq a^a$,
(ii) *if* $a > 1$ *and* $u \geqq a^{2a}$ *then* $\mathcal{Q}(a, u)$.

In fact, we may take for $\mathcal{Q}(a, u)$ the relation between a and u which holds if and only if there are x and y such that

$$\begin{aligned} &x^2 - (a^2 - 1)(a - 1)^2y^2 = 1, \\ &x > 1, \\ &u \geqq ax. \end{aligned} \tag{11.5}$$

Proof: Suppose $\mathcal{Q}(a, u)$ holds. Since $x > 1$, $a > 1$. Hence there is an $n > 0$ such that $x = a_n$ and $(a - 1)y = a'_n$. By Lemma 2, $n \equiv 0 \pmod{a - 1}$. Therefore $n \geqq a - 1$. Hence $u \geqq a \cdot a_{a-1}$. But $a_n \geqq a^n$ by (11.1), so $u \geqq a^a$.

Now for $a > 1$, the conditions (11.5) will be satisfied by $x = a_{a-1}$, $y = a'_{a-1}/(a - 1)$, and $u \geqq a \cdot a_{a-1}$. Since a_n is the rational part of $(a + \sqrt{a^2 - 1})^n$, we see that $a_n \leqq (2a)^n \leqq a^{2n}$. Hence $a \cdot a_{a-1} \leqq a \cdot a^{2a-2} < a^{2a}$. Therefore, if $u \geqq a^{2a}$ and $a > 1$ then $\mathcal{Q}(a, u)$.

THEOREM: *The relation* $r = s^t$ *is diophantine if any infinite set of primes is diophantine.*

Proof: We shall show that $r = s^t$ with $s > 0$ if and only if there are natural numbers a, x, y, and p such that

$$\begin{aligned} &s > 0,\ a > 1, \\ &\mathcal{Q}(s + t + 1, p), \\ &p \text{ is a prime}, \\ &p \mid 2as - s^2 - 1,\ p - 1 \mid a - 1. \\ &x^2 - (a^2 - 1)y^2 = 1, \\ &y \equiv t \pmod{a - 1}, \\ &\operatorname{Rem}(x - (a - s)y, p) = r, \end{aligned} \tag{11.6}$$

Here $\mathcal{Q}$ is the relation of Lemma 3 and all the conditions are clearly diophantine except for "p is a prime." In order to obtain the theorem, we need to prove slightly more. It is sufficient to show that if (11.6) is satisfied, then $r = s^t$, and if $r = s^t$ and $s > 0$, then (11.6) can be satisfied for any sufficiently large prime p. Hence we could replace the condition "p is a prime" by the condition $p \in \mathcal{P}$ where $\mathcal{P}$ is any infinite diophantine set of primes.

Of course, if $r = s^t \wedge s > 0$ is diophantine, then $r = s^t$ is diophantine.

Suppose (11.6) holds. Then $s^t < p$, $0 < s < p$, and for some n, $x = a_n$ and $y = a'_n$. So by Lemma 2,

$$x - (a - s)y \equiv s^n \pmod{2as - s^2 - 1}$$

and

$$y \equiv n \pmod{a - 1}.$$

Hence by (11.6),

$$r \equiv s^n \pmod{p}$$

and

$$t \equiv n \pmod{p - 1}.$$

Since $s^{p-1} \equiv 1 \pmod{p}$ by Fermat's theorem,

$$r \equiv s^n \equiv s^t \pmod{p}.$$

But $s^t < p$ and $r < p$, so $r = s^t$.

Now suppose $r = s^t$ and $s > 0$. Let p be a prime so large that $\mathfrak{Q}(s + t + 1, p)$ holds. We need to choose $a > 1$ so that

$$\begin{aligned} p - 1 &\mid a - 1, \\ p &\mid 2as - s^2 - 1. \end{aligned} \tag{11.7}$$

If a satisfies (11.7), then $x = a_t$ and $y = a'_t$ will satisfy the other conditions of (11.6) and we are through. But (11.7) amounts to

$$\begin{aligned} a &\equiv 1 \pmod{p - 1}, \\ 2sa &\equiv s^2 + 1 \pmod{p}, \end{aligned}$$

and since $(2s, p) = 1$, we can solve these congruences simultaneously for a. Hence all the conditions of (11.6) can be satisfied.

As a consequence of this theorem, we obtain the result that *if an infinite set of primes is diophantine, then every recursively enumerable set is diophantine and there is no general method to tell if an arbitrary diophantine equation has a solution.*

Remark: In order to satisfy (11.6) for a given r, s, t with $r = s^t$, it is sufficient to take for p any sufficiently large prime. Hence we can simultaneously define a finite set of relations

$$r_1 = s_1^{t_1}, \cdots, r_k = s_k^{t_k}$$

in terms of a single prime. We need only repeat the conditions of (11.6) with $r_i, s_i, t_i, a_i, x_i, y_i$ replacing r, s, t, a, x, y for $i = 1, \cdots, k$, leaving p fixed throughout. Hence *to every exponential diophantine set* $\mathcal{S}$, *i.e. every recursively enumerable set* $\mathcal{S}$, *there is a diophantine equation* $P(x, p, y_1, \cdots) = 0$ *which has a solution with* p *prime for a given* x *if and only if* $x \in \mathcal{S}$. This argument is similar to one used by Davis.

COROLLARY: *There is no general method of deciding whether or not*

$$\text{Range } A(x_1, \cdots) \subseteqq \text{Range } ((u + (v + 2)(w + 2))^2 + u)$$

where A *is an arbitrary polynomial with positive integer coefficients.*†

Proof: From the above remark, there is no general method to tell whether or not

$$F(p, y_1, \cdots) = G(p, y_1, \cdots)$$

has a solution with p prime where F and G are arbitrary polynomials with positive integer coefficients. Hence there is no method to tell whether the polynomial

$$(z + 2)(1 + (F(z + 2, y_1, \cdots) - G(z + 2, y_1, \cdots))^2)$$

represents a prime, i.e., assumes a prime number as value.

Let $T(x, y) = (x + y)^2 + x$, so that $T(x, y) = T(u, v)$ implies that $x = u$ and $y = v$. Then the polynomial $A(z, y_1, \cdots)$ given by

$$T((z + 2) \cdot 2FG, (z + 2)(1 + (F - G)^2))$$

has positive coefficients. Furthermore,

$$\text{Range } A(z, y_1, \cdots) \subseteqq \text{Range } T(u, (v + 2)(w + 2))$$

if and only if $(z + 2)(1 + (F - G)^2)$ does not represent a prime. Hence there is no general method of telling whether or not

$$\text{Range } A(z, y_1, \cdots) \subseteqq \text{Range } T(u, (v + 2)(w + 2)),$$

where A is an arbitrary polynomial with positive coefficients.

Next we use the same lemmas and a similar argument to prove another theorem.

† **Open problem:** Is there a decision method to tell whether or not an arbitrary polynomial with positive integer coefficients represents all natural numbers?

THEOREM: *If there is a diophantine relation $\mathcal{S}(p, q)$ such that*

(11.8) *if $\mathcal{S}(p, q)$, then $p > 1$ and $q \leqq p^p$,*

and

(11.9) *for every k, there are p and q with $\mathcal{S}(p, q)$ and $q > p^k$,*

then the relation $r = s^t$ is diophantine.

Proof: We shall show that $r = s^t$ with $s, t > 0$ if and only if there are natural numbers a, x, y, and z such that

$$\begin{aligned}
&s > 0, \quad t > 0,\\
&\mathcal{Q}(s + t + 1, 2as - s^2 - 1),\\
&\mathcal{S}(a, z),\\
&x < z, \quad y > 0, \quad x^2 - (a^2 - 1)y^2 = 1,\\
&\text{Rem } (y, a - 1) = t,\\
&\text{Rem } (x - (a - s)y, 2as - s^2 - 1) = r.
\end{aligned} \tag{11.10}$$

Suppose (12.10) is satisfied. Then $s^t < 2as - s^2 - 1$, $x < a^a$ and $y > 0$. Hence $x = a_n$ and $y = a'_n$ for some positive $n < a$. Also Rem $(y, a - 1) = t \equiv n \pmod{a - 1}$, so $n = t$. Finally, Rem $(x - y(a - s), 2as - s^2 - 1) = r \equiv s^t (\text{mod } 2as - s^2 - 1)$ so $r = s^t$.

Suppose $r = s^t$ with $s, t > 0$. The conditions (11.8) and (11.9) on $\mathcal{S}$ insure that for every k there are infinitely many p and q with $\mathcal{S}(p, q)$ and $q > p^k$. Hence we can choose a so large that

$$\begin{aligned}
&\mathcal{Q}(s + t + 1, 2as - s^2 - 1),\\
&t < a - 1,
\end{aligned}$$

and such that for some z, $\mathcal{S}(a, z)$ and $z > a^{2t}$. Then for this choice of a, $z > a_t$. Hence if we take $x = a_t$ and $y = a'_t$, all conditions of (11.10) will be satisfied.

THEOREM: *If there is a diophantine relation $\mathcal{R}(p, q)$ such that*

(11.11) *there is some n such that if $\mathcal{R}(p, q)$ then $q \leqq p * n$*

and

(11.12) *for every k, there are p and q with $\mathcal{R}(p, q)$, $p > 1$, and $q > p^k$,* *then there is a diophantine relation* $\mathcal{S}(p, q)$ *which satisfies* (11.8) *and* (11.9).

Proof: Case A. Suppose $\mathcal{R}(u, v)$ implies $v \leqq u^{mu}$ for some m. Let $\mathcal{S}(p, q)$ be the diophantine relation which holds if and only if there is a v such that

$$(11.13) \qquad p > 1, \quad \mathcal{R}(p, v), \quad \text{and} \quad q^m \leqq v.$$

If $\mathcal{S}(p, q)$, then for some v, $q^m \leqq v \leqq p^{mp}$ so $q \leqq p^p$. Hence $\mathcal{S}$ satisfies (11.8). On the other hand, for every k there is a $p > 1$ and a v so that $\mathcal{R}(p, v)$, $v > p^{2mk}$, and hence $\mathcal{S}(p, p^{2k})$. Since $p > 1$, $p^{2k} > p^k > p^0$ for $k > 0$. Therefore $\mathcal{S}$ satisfies (11.9).

Case B. Suppose $\mathcal{R}(u, v)$ implies $v \leqq u * n$ for some $n > 2$ and there is no m such that $\mathcal{R}(u, v)$ implies $v \leqq u^{mu}$. Let $\mathcal{R}_1(u, v)$ if and only if there is a natural number a such that $\mathcal{Q}(a, u)$ and $\mathcal{R}(a, v)$. Hence if $\mathcal{R}_1(u, v)$ holds, there is an a with

$$\begin{aligned} a^a &\leqq u, \\ v \leqq a * n \leqq a^a * (n-1) &\leqq u * (n-1). \end{aligned}$$

Furthermore, for every k, there is an $a > 1$ and v with $\mathcal{R}(a, v)$ and $v > a^{2ka}$. By Lemma 3, $\mathcal{Q}(a, a^{2a})$. Hence for $u = a^{2a}$, we have $\mathcal{R}_1(u, v)$ and $v > u^k$. Thus we obtain the theorem, since we can continue to apply Case B until either n is reduced to 2 or we obtain a relation which falls under Case A.

Putting the last two theorems together with the fact that every recursively enumerable relation is exponential diophantine, we obtain the following.

THEOREM: *If there is a diophantine relation* $\mathcal{R}(p, q)$ *which satisfies* (11.11) *and* (11.12), *then every recursively enumerable relation is diophantine.*

12. CONCLUSIONS

Exactly one of the following two statements is true:

I. Every recursively enumerable relation is diophantine.

II. For every diophantine equation $P(p, q, x_1, \cdots, x_k) = 0$ either (1) for every n there is a solution with $q \geqq p * n$ or (2) there is an m such that every solution satisfies $q \leqq p^m$.

If I is true, then all the familiar number-theoretic sets such as the set of primes would be diophantine and as in the case of exponential diophantine equations, there could be no method of telling whether an arbitrary diophantine equation is solvable.

On the other hand, if II is true, there would be rather surprising bounds on solutions of diophantine equations. We see that (1) holds whenever there is a p such that there are infinitely many q for which $P(p, q, x_1, \cdots) = 0$ has a solution or if there is a solution with $p = 0$ or 1 and $q > 1$. Does (1) hold in any other case? There is no example known. Hence it may be that if for all $p > 1$ there are at most a finite number of q's for which $P = 0$ has a solution, then there is an m such that all solutions with $p > 1$ satisfy $q < p^m$. If this is the case, then a positive solution to the decision problem for diophantine equations would appear more likely although by no means assured. We would still need a test to tell if for some p there are infinitely many q's for which the given equation has a solution and some bound on m in terms of the coefficients of the equation.

Is it I or II that is true? We shall probably find out by disproving one of them. For example, if we could show that some recursively enumerable set is not diophantine, then I is false. Or it would be sufficient to show that the class of recursively enumerable sets differs in some regard from the class of diophantine sets. On the other hand, to disprove II, all we need is a single diophantive equation which is a counter-example. Davis and Putnam have recently made progress in this direction. (See [2], [3].)

We now put these results together.

THEOREM: *Consider diophantine equations $P(p, q, x_1, \cdots) = 0$ such that to every p there are at most a finite number of values of q for which $P = 0$ is solvable. If to every such equation $P = 0$, there is an n such that every solution of $P = 0$ with $p > 1$ satisfies $q \leqq p * n$, then to every such equation there is an n' such that each solution with $p > 1$ satisfies $q \leqq p^{n'}$.*

Proof: Suppose some diophantine relation $\mathcal{R}(p, q)$ satisfies (11.11) and (11.12). Then every recursively enumerable relation is diophantine. In particular, the relation given by $q = p * p$ is diophantine. Hence there is an equation $P_1(p, q, x_1, \cdots) = 0$ which is solvable for exactly one q for each p and for which there is no n such that all solutions with $p > 1$ of $P_1 = 0$ satisfy $q \leqq p * n$. So the theorem is true in this case. On the other hand, if all the equations considered define relations which satisfy (11.11) then the relations do not satisfy (11.12), i.e., to each of these equations, there is an n' such that every solution with $p > 1$ satisfies $q \leqq p^{n'}$.

BIBLIOGRAPHY

1. Davis, Martin, *Computability and Unsolvability.* McGraw-Hill, New York, 1958.
2. ———, "Applications of recursive function theory to number theory," *Recursive Function Theory, Proceedings of Symposia in Pure Mathematics, Volume V. Amer. Math. Soc.* (1962), 135–138.
3. ———, "Unsolvable problems: a review," *Proceedings of the Symposium on Mathematical Theory of Automata, Microwave Res. Inst. Symposia Series,* Vol. 12, Polytechnic Press, Brooklyn, 1963, 15–22.
4. ———; Putnam, Hilary; and Robinson, Julia, "The decision problem for exponential diophantine equations," *Ann. of Math.,* 74 (1961), 425–436.
5. Gödel, Kurt, "Über formal unentscheidbare Sätze der Principia Mathematica und verwandter Systeme I," *Monatsh. Math. und Physik,* 38 (1931), 173–198. For an English translation of this article, see Kurt Gödel, *On Formally Undecidable Propositions,* Basic Books, 1962.
6. Grzegorczyk, Andrzej, *Fonctions Récursives.* Gauthier-Villars, Paris, 1961.
7. Hermes, Hans, *Aufzählbarkeit, Entscheidbarkeit, Berechenbarkeit.* Springer-Verlag, Berlin, 1961.
8. Kleene, Stephen Cole, *Introduction to Metamathematics.* Van Nostrand, Princeton, N.J., 1952. Chapters 11–13.

9. LeVeque, William J., *Elementary Theory of Numbers.* Addison-Wesley, Reading, Mass., 1962.

10. Mendelson, Elliott, *Introduction to Mathematical Logic.* Van Nostrand, Princeton, N.J., 1964. Chapter 5.

11. Post, Emil, "Recursively enumerable sets of positive integers and their decision problems," *Bull. Amer. Math. Soc.* 50 (1944), 284–316.

12. Putnam, Hilary, "An unsolvable problem in number theory," *J. Symb. Logic,* 25 (1960), 220–232.

13. Robinson, Julia, "Existential definability in arithmetic," *Trans. Amer. Math. Soc.,* 72 (1952), 437–449.

COMPUTER TECHNOLOGY APPLIED TO THE THEORY OF NUMBERS

D. H. Lehmer

The era of the automatic digital computer is nearing the end of its second decade. By this time a fair proportion of number theorists are becoming aware of its existence and are at least wondering what the commercial electronic computer might be able to do for them. Many number theorists have already experienced one or more instances of aid from this quarter.

Actually, the application of electronics to number theory began more than 30 years ago. By now enough experience has been gained and enough standardization achieved to enable one to give a structured account of what can be accomplished in this field today. We do not propose to give a chronological history of such applications. Instead, we confine ourselves to a topically arranged discussion of what can be done with only standard equipment and at what costs in terms of machine time.

Before entering our subject in detail, we insert here a few simple remarks in reply to the question: Why might a number theorist want to add pulse circuitry to pencil and paper? These remarks are arranged in order of increasing machine involvement.

The history of the theory of numbers abounds with conjectures or announcements of theorems without proof. These statements contain in each case at least one parameter n which ordinarily takes on infinitely many integer values. By using a machine, one's ability to explore is enormously enhanced and so one may be fortunate enough to discover at least one value of n for which the statement is patently false. This settles the question of the truth value of the statement, once and for all. Many future man hours of fruitless attempts to establish the truth of the proposition are thus saved. It makes no difference how this counter-example was discovered. It is not even necessary to give due credit to a mere machine now that anyone can easily verify for himself that the proposition is false. However, in the back of one's mind there should be some gratitude for the machine that has opened our eyes to a new situation in the complex universe of the integers.

In the previous situation one may not be fortunate enough to discover a counter-example. The more the machine succeeds in verifying the truths of the proposition in special cases of n, the more plausible the truth of the proposition becomes, although plausibility is apt to be a nonlinear function of n as for example in the proposition: Every even number (>2) less than $2n$ is the sum of two primes. Although we cannot obtain a proof by this device, there is the possibility that a careful analysis of many cases will reveal "why" the proposition is generally true and thus the machine has helped by leading to the idea of the proof. If properly directed, the machine itself can undertake this analysis.

Most of the important classical theorems in number theory were discovered as a by-product of the production and inspection of tables. These were constructed by hand. The modern machine can produce tables with speed and reliability many orders of magnitude greater than what is humanly possible. Not only is the publication of such tables impossible; even the inspection is often well beyond human capability. It soon becomes apparent that it should be the machine's responsibility to make this inspection, with, of course, a little sound advice from the programmer.

The verification of all of a given large number of statements is an example of what is called exhaustive search. This kind of activity is not well suited to humans, not only because of the large number of different things to do, but also because of the complexities of the relationships or arrangements of the things themselves. Fortunately, the modern digital computer is not only very fast but also very adroit in handling combinatorial complexities when carefully programmed. At the end of the run, the machine announces, in words introduced in advance by the programmer, that all cases have been considered and verified. Having entrusted the machine with this much responsibility, one must have faith that it performed all its instructions correctly. Those of little faith will ask: Is this really mathematics since it cannot be done at the blackboard?

It is but a simple step from the preceding instance to one kind of mechanical "theorem proving." This is not the kind of machine proof with a "look, no hands!" point of view in which the machine starts from the postulates and proves a well known elementary theorem, simulating well established heuristic procedures in its search for a proof. Rather it is a man-machine cooperative endeavor in which the man furnishes the best information he has as to the kind of proof most likely to succeed and the machine attempts to carry out all the steps by exhaustive search. The efficient use of the machine demands that the number of steps to be carried out be very large but finite and that the relations between the steps be combinatorily complicated. Most proofs don't meet these specifications. Those that do are proofs by separation of cases. So far they have been the proofs of theorems in number theory, previously unproved. Since the machine consists of a finite number of circuit elements, each capable of being in a finite number of stable states separated by finite time intervals, and since the machine runs for a finite number of minutes, it is clear that in at least one of its many steps of the proof the machine is dealing with an infinite number of values of the parameter n.

We turn now to some preliminary consideration of the equipment available to the number theorist in typical university computing centers. This equipment is of two sorts: hardware and software.

On the hardware side one finds a variety of computing systems. They can be classified roughly into small and large installations, although this distinction is becoming more blurred with time. Small machines have fewer words of storage and slower rates of operation and, for number theory problems, are about two orders of magnitude less effective than large machines. If one used either machine full time the larger one would accomplish about a hundred times more work. Of course, one does not operate this way. Many problems are quite short and not very demanding of storage space. Here it makes little difference whether the running time is five minutes or five seconds. If a large problem is programmed on a standby basis, so that the machine gives attention to the problem whenever it would otherwise stand idle, a great deal of work can be put through a small machine in a week's time.

Many small machines are decimal rather than binary. These are decidedly less satisfactory for number theory and combinatorial problems since one has no access to the separate binary bits of a machine word. However, in some problems in which the number ten is basically involved a decimal machine can be superior to an otherwise comparable binary one.

By "software" is meant the system of executive programs used by the machine operators and programmers to put jobs through the machine. These programs include routines for assembling, compiling, monitoring, and translating the given instructions from the programmer's language to that of the machine. The chief object of software is to put through as many short jobs as possible in a day's time, often as many as 500 or 1,000. This laudable purpose is important for the businesslike operation of a computer for an industry or a university that behaves like an industry but it imposes restraints on any particular user, especially a low priority user like the number theorist.

The language problem is a case in point. Languages like ALGOL and FORTRAN stand between the user and the machine, to "help him communicate." These languages have been designed by people who know best what the user wants to compute. Of course, the contemplated user is never a number theorist. For example, in FORTRAN II all positive integers are less than 32,768 and multiplication is only approximate. In ALGOL no account is taken of the structure of the machine word. Even the present writer does not know best what a number theorist is going to compute, but it doesn't take years of experience to discover that many number theory problems are made feasible only by having access to the full repertory of the machine's instructions in the machine's own language.

This tyranny of the master language has been felt in other parts of computing, such as the "list processing" branch and has led to the inauguration and development of splinter languages, some of which are applicable to number theory. At present writing, there is no sign of a number theory language emerging, no doubt because of the economics of the situation. Languages cost real money. However, the needs of the number theorist are pretty well met by a package of much used subroutines written in machine language. Such a package is described in Section 9.

A fair-sized number theory program, containing from 1,000 to 10,000 separate machine instructions, will ordinarily be made up of several subroutines. Before one contemplates the practical cost of executing such a program, one must have some idea of the total amount of time the machine will spend in each of the subroutines. Some routines require a fixed time for execution independent of the data involved. Others, and these are likely to be the crucial ones, are time dependent upon one or more parameters. We say that a subroutine or an algorithm is of order $f(n)$ in the parameter n if the time of execution is less than a constant multiple of $f(n)$. This constant, of course, depends on the speed of the machine and also on the "tightness" of the programming. Ignoring these important considerations, it is the function $f(n)$ which may well govern the feasibility of the entire program.

We describe in what follows a number of basic algorithms of different orders beginning with those of order $\log n$.

ALGORITHMS OF ORDER LOG n

1. THE GREATEST COMMON DIVISOR ALGORITHM

Historically the first statement as to the order of an algorithm was made in 1844 by Lamé. This is his well-known theorem to the effect that the number of divisions necessary to find the greatest common divisor (m, n) of two positive integers $m > n$ does not exceed five times the number of decimal digits of n. That the factor 5 can be reduced to 3 by using the nearest integer algorithm rather than the standard Euclidean algorithm considered by Lamé is of secondary interest. The important fact is that the average execution time of any G.C.D. algorithm will vary directly as the logarithm is its smaller parameter.

Since greatest common divisor algorithms and regular continued fractions are quite fundamental, it may be worth while to make a few comments about them. It is traditional to describe the Euclidean algorithm in terms of a finite sequence of relations of the form

$$a_{i-1} = q_i a_i + a_{i+1}, \quad 0 \leqq a_{i+1} < a_i. \tag{1.1}$$

If one is interested only in the greatest common divisor of a_0 and a_1, there is no need to retain the partial quotients q_i nor, indeed, any of the intermediate remainders a_i. By the metric theory of continued fractions, 41 per cent of the q's are equal to 1 and 17 per cent of the rest are equal to 2. [The probability that a particular q is $\geqq m$ is $\{\log(1 + m^{-1})\}/(\log 2)$.] This means that more than half the time, division, a relatively expensive instruction, can be replaced by one or two subtractions. This fact can be important when one is dealing with multiprecise integers (integers occupying two or more machine words) if one is coding for speed.

The following simple procedure for obtaining $(m, n) = d$ where $m > n$, involves no division at all. It is a good example of elimination of the fussy subscripts of (1.1). In this, and in future formal procedures, one goes from Step k to Step $k + 1$ unless ordered to go elsewhere in Step k. The notation $a \to b$ means a replaces b.

Basic Nearest Integer G.C.D. Procedure

$$(m, n) = d, \quad (m > n > 0)$$

1. $m \rightarrow s$
2. $s - n \rightarrow s$
3. If $s > n$ go to Step 2.
4. If $s = n$ go to Step 11.
5. $n \rightarrow m$
6. If $2s \leqq n$ go to Step 9.
7. $n - s \rightarrow n$
8. Go to Step 1.
9. $s \rightarrow n$
10. Go to Step 1.
11. $n \rightarrow d$
12. Halt.

In the above, s represents an auxiliary storage variable. The difficulty with this simple procedure is that occasionally a good deal of time may be spent in Step 2 and 3 in the attrition of s by successive subtractions of n. These are the cases when q_i is large and one should use division. To combine the best features of both the classical Euclidean algorithm and the above procedure one has only to set up a tally counting the number of times Step 3 leads to Step 2. Whenever this tally exceeds a certain limit determined by the relative costs of division and subtraction, the machine will proceed to divide instead of continuing to subtract. This same tally will be useful (and indeed necessary) in those modifications of the above procedure for applications of the algorithm that use the partial quotients q_i, for example in the continued fraction expansion of m/n, the solution of the Diophantine equation

$$mx + ny = d,$$

the linear congruence

$$ny \equiv d \pmod{m},$$

and Chinese remainder problems.

2. THE JACOBI SYMBOL

A somewhat similar algorithm of order log n is that for evaluating the Jacobi symbol. The reciprocity law

$$\left(\frac{n}{m}\right) = (-1)^{(m-1)(n-1)/4}\left(\frac{m}{n}\right), \quad (m, n \text{ odd})$$

and the reduction formula

$$\left(\frac{m}{n}\right) = \left(\frac{m-n}{n}\right)$$

lead to a procedure involving subtraction and interchanging in much the same spirit as the G.C.D. algorithm.

3. POWER ALGORITHM

An entirely different basic algorithm of order log n is that for finding powers with respect to a modulus or, more explicitly, finding r_n in

$$b^n \equiv r_n \pmod{m}, \quad 0 \leqq r_n < m,$$

given b, n, and m. If we were to use the simple recurrence

$$r_{k+1} \equiv b^{k+1} \equiv br_k \pmod{m},$$

the resulting algorithm would be of order n, not log n. Instead we use either

$$r_{2k} \equiv b^{2k} \equiv r_k^2 \pmod{m} \tag{3.1}$$

or

$$r_{2k+1} \equiv b^{2k+1} \equiv br_k^2 \pmod{m} \tag{3.2}$$

and thus build up the nth power by successive squaring with an occasional multiplication by b whenever the integer n has a unit digit in its binary representation. Thus, the total number of multiplications is less than a constant times log n. This method of constructing b^n often uses only the minimum possible number of multiplications. (The optimum procedure is a complicated function of n.) For large n and for b a single machine word the cost of the algorithm is practically independent of b.

The following procedure makes the above method precise.

POWER ALGORITHM PROCEDURE

1. If $n = 0$ go to Step 20.
2. $0 \rightarrow T$
3. If n is odd go to Step 7.
4. $n/2 \rightarrow n$
5. $1 + T \rightarrow T$
6. Go to Step 3.
7. $n + n \rightarrow n$
8. If no overflow go to Step 7.
9. $b \pmod{m} \rightarrow r$
10. If $n = 0$ go to Step 16.
11. $r^2 \pmod{m} \rightarrow r$
12. $n + n \rightarrow n$
13. If no overflow go to Step 10.
14. $br \pmod{m} \rightarrow r$
15. Go to Step 10.
16. If $T = 0$ go to Step 21.
17. $r^2 \pmod{m} \rightarrow r$
18. $T - 1 \rightarrow T$
19. Go to Step 16.
20. $1 \rightarrow r$
21. Halt.

This procedure uses intentional overflow to control itself. In Steps 7 and 12 (which are, in general, multiprecise additions carried out by a special subroutine) we assume that the overflow digit is lost. For example, with a ten digit decimal accumulator we would have, for instance,

$$6{,}528{,}000{,}000 + 6{,}528{,}000{,}000 = 3{,}056{,}000{,}000$$

with the overflow indicator set. For a binary machine these Steps 7 and 12 could be replaced by left shifts of 1. In Steps 9, 11, 14, and 17, reduction modulo m occurs. This, in general, requires a multiprecise division subroutine except when m is very close to a power of 2, for a binary machine, or a power of 10 for a decimal machine.

The fact that the power algorithm is of order log n has several important consequences, as we shall now see.

4. POWER CHARACTER DETERMINATION

Let $k > 2$ be an integer and let $p = kn + 1$ be a prime with a primitive root g. Let b be a given integer not divisible by p. The problem with which we are concerned is to find the kth power character of b modulo p with respect to g. That is, we wish to determine the integer β, $(0 \leqq \beta < k)$ such that

$$\text{Ind}_g\, b \equiv \beta \pmod{k}.$$

Often one is interested only in whether $\beta = 0$, that is, whether b is a kth power residue of p. In this case, the choice of g is immaterial and we have Euler's criterion

$$\beta = 0 \Leftrightarrow b^n \equiv 1 \pmod{p}.$$

The determination of $b^n \pmod{p}$ is a problem of order log n and hence of order log p. Other criteria for kth power residuacity exist for small k. These are indeed remarkable but their applications involve algorithms of order $p^{1/2}$ and hence are more costly.

Returning to the problem of determining β, we note that the number

$$\epsilon \equiv g^n \pmod{p}$$

is a primitive kth root of unity modulo p, that is, if

$$\epsilon^i \equiv r_i \pmod{p}, \quad (i = 0(1)k - 1),$$

the r_i are distinct and $\epsilon^k \equiv 1 \pmod{p}$. These k values r_i can be computed and stored for future reference. Now let

$$\text{Ind}_g\, b = \beta + \alpha k.$$

Then

$$b^n \equiv g^{n\beta+\alpha kn} \equiv g^{n\beta+\alpha(p-1)} \equiv g^{n\beta} \equiv \epsilon^\beta \equiv r_\beta \pmod{p}.$$

Hence b^n is congruent modulo p to one of the k stored values of r. The number β can hence be determined by the machine by "inverse table look up." The whole procedure involves a cost proportional to log p.

5. PRIMITIVE ROOT DETERMINATION

In the previous section we have postulated a primitive root g of p. Although it is used only in determining ϵ, it is of some interest to ask in general how expensive is the determination of a primitive root of a large prime p. Gauss gave a recursive algorithm for determining a primitive root, but it is of order p. If one knows the prime factors of $p - 1$, then a primitive root can soon be found by applying the following criterion:

If $g^{(p-1)/q} \equiv 1 \pmod{p}$ *holds for none of the prime factors* q *of* $p - 1$, *then* g *is a primitive root of* p.

The application of this criterion is an effort of order $(\log p)^2/\log \log p$ at most. One starts out boldly substituting for g the early quadratic nonresidues of p and expects to find a primitive root almost at once. In spite of an extensive literature on the least primitive root problem, no realistic results have thus far been found. Although there are primes whose least quadratic nonresidue and hence whose least primitive root exceeds any preassigned number, there is a strong heuristic argument to the effect that any quadratic nonresidue of p has a probability of nearly $\frac{3}{4}$ of being a primitive root of p. In practice, therefore, one has no real difficulty in discovering a primitive root of p once the prime factors of $p - 1$ are known. The problem of factoring $p - 1$ is, as we shall see, a problem of order $p^{1/2}$.

6. TESTS FOR PRIMALITY

In recent years many exceedingly large primes have been identified. This reflects the fact that the mere determination of the prime or composite nature of a number N is, with a little luck, a problem of order log N. This comes about because the power algorithm is of order log N. The part that is left to luck is the discovery of a more or less complete factorization of $N - 1$ or $N + 1$, as we are now about to explain.

The simplest case is that in which all the prime factors of $N - 1$ are known. We then apply our algorithm to see if the hypothesis of the following modified converse of Fermat's theorem is satisfied.

THEOREM 1: *If for some choice of* b

$$b^{N-1} \equiv 1 \pmod{N} \tag{6.1}$$

while

$$b^{(N-1)/q} \not\equiv 1 \pmod{N}$$

holds for all of the prime factors q *of* $N - 1$, *then* N *is a prime. In case* (6.1) *fails to hold for some* b, *then* N *is composite.*

This theorem has been in use since 1925. Incidentally, once N has been proved prime by this criterion, b is a primitive root of N.

If one is less lucky, not all the prime factors of $N - 1$ will be known. In this case, one or more applications of the following theorem may suffice to restrict the possible prime factors p of N to such an extent that their complete nonexistence can be established.

THEOREM 2: *Let* q *be a prime factor of* $N - 1 = q^\alpha m$. *If for some choice of* b

$$b^{N-1} \equiv 1 \pmod{N}$$

while $b^{(N-1)/q} - 1$ *is relatively prime to* N, *then every prime factor* p *of* N *is of the form* $p = q^\alpha x + 1$.

If one has no luck with trying to factor $N - 1$, one can deal instead with $N + 1$. This leads to a slightly more elaborate algorithm, based on the theory of Lucas' functions U_n, V_n, and designed for computer application. The method is only three times as expensive as the previous one. The counterpart of Theorem 1 is the following.

THEOREM 3: *Let* N *be an odd integer. Let* Q *be chosen so that the Jacobi symbols*

$$\left(\frac{Q}{N}\right) \quad \text{and} \quad \left(\frac{1-4Q}{N}\right)$$

are both negative. Let α, β *be the roots of*

$$x^2 - x + Q = 0$$

and let

$$V_n = \alpha^n + \beta^n.$$

Next suppose that

$$V_{(N+1)/2} \equiv 0 \pmod N \tag{6.2}$$

and that

$$V_{2(N+1)/q} \not\equiv 2Q^{(N+1)/q} \pmod N \tag{6.3}$$

holds for all of the odd prime factors q of $N + 1$. Then N is a prime. If on the other hand (6.2) *fails to hold, then N is composite.*

Before sketching the proof of this theorem, we wish to show that its application is an algorithm of order $\log N$. First it is easy to verify that the integers V_n satisfy

$$V_n = V_{n-1} - Q \cdot V_{n-2}. \tag{6.4}$$

If one relied on this recurrence alone the verification of (6.2) would be a calculation of order N. Instead, we use the duplication formula

$$V_{2m} = V_m^2 - 2Q^m. \tag{6.5}$$

Starting with two consecutive V's, namely

$$V_0 = 2 \quad \text{and} \quad V_1 = \alpha + \beta = 1,$$

we build a succession of pairs of consecutive V's. Given (V_k, V_{k+1}) we have by (6.5)

$$V_{2k} = V_k^2 - 2Q^k,$$
$$V_{2k+2} = V_{k+1}^2 - 2Q^{k+1}.$$

Hence by (6.4) we can obtain either

$$(V_{2k}, V_{2k+1}) \quad \text{or} \quad (V_{2k+1}, V_{2k+2})$$

corresponding to the choice of (3.1) or (3.2).

As in the case of the power algorithm procedure, using intentional overflow, one can make the machine program its own calculation of $V_n \pmod N$ in only $O(\log N)$ steps. Similarly, as soon as one finds (V_{w-1}, V_w), where w is the largest odd divisor of $n = 2^\lambda w$ one can proceed directly to V_n by λ applications of the duplication formula (6.5).

We now proceed to sketch a proof of Theorem 3.

Proof: Let

$$U_n = \frac{(\alpha^n - \beta^n)}{(\alpha - \beta)}$$

so that

$$U_{2k} = U_k V_k \tag{6.6}$$

and

$$(1 - 4Q)U_n^2 = V_{2n} - 2Q^n = V_n^2 - 4Q^n. \tag{6.7}$$

If, for some integer $m > 1$, m divides U_k but not U_h for $0 < h < k$, we say that k is the rank of apparition of m and write

$$\rho(m) = k.$$

It is easily seen that

$$m \mid U_n \Leftrightarrow \rho(m) \mid n \tag{6.8}$$

and it is known that if p is a prime,

$$\rho(p^\alpha) \mid p^{\alpha-1}\left(p - \left(\frac{1 - 4Q}{p}\right)\right). \tag{6.9}$$

With these preliminaries, let

$$N = p_1^{\alpha_1} p_2^{\alpha_2} \cdots p_t^{\alpha_t}, \quad (p_i \geq 3)$$

be the canonical factorization of the odd number N.

Our first task is to show that if (6.2) and (6.3) hold, then $t = 1$ and $\alpha_1 = 1$, so that N is a prime. By (6.6)

$$U_{N+1} = U_{(N+1)/2} V_{(N+1)/2} \equiv 0 \pmod{N}.$$

Hence from (6.8)

$$\rho(N) \mid N + 1.$$

For each odd prime factor q of $N + 1$, in view of (6.7) and (6.3),

$$(1 - 4Q)U_{(N+1)/q}^2 = V_{2(N+1)/q} - 2Q^{(N+1)/q} \not\equiv 0 \pmod{N}.$$

Hence $\rho(N)$ contains the same highest power of q that $N + 1$ does. By (6.4) and (6.7) any odd prime factor common to U_n and V_n would divide V_{n-1} and hence would divide $V_1 = 1$. In view of (6.2) then, N is prime to $U_{(N+1)/2}$. Therefore $\rho(N)$ and $N + 1$ contain the same highest power of 2. Therefore $\rho(N) = N + 1$. On the other hand, $\rho(N)$ divides the least common multiple of $\rho(p_i^{\alpha_i})$ $(i = 1(1)t)$ and hence by (6.9), $\rho(N)$ divides

$$m = 2 \prod_{i=1}^{t} p^{\alpha_{i-1}} \left\{ p_i - \left(\frac{1+4Q}{2} \right) \right\} / 2$$
$$\leq 2N \prod_{i=1}^{t} \frac{1}{2} \left(1 + \frac{1}{p_i} \right) \leq 2N \left(\frac{2}{3} \right)^t.$$

Hence

$$N + 1 = \rho(N) \leq m \leq 2N(\tfrac{2}{3})^t.$$

This forces t to be 1 and m to be $p^{\alpha-1}(p \pm 1)$. This can be divisible by $\rho(N) = N + 1 = 1 + p^\alpha$ only if $\alpha = 1$. Thus N is the prime p.

Suppose on the other hand (6.2) fails to hold and nevertheless N is a prime. By (6.9)

$$\rho(N) \mid N - \left(\frac{1-4Q}{N} \right) = N + 1.$$

Hence N divides $U_{N+1} = U_{(N+1)/2} V_{(N+1)/2}$ and since (6.2) does not hold, N divides $U_{(N+1)/2}$. But in this event by (6.7)

$$(V_{(N+1)/2})^2 \equiv 4Q^{(N+1)/2} \equiv 4Q \left(\frac{Q}{N} \right) \equiv -4Q \pmod N.$$

This means that, by (6.5),

$$\begin{aligned} V_{N+1} &= V^2_{(N+1)/2} - 2Q^{(N+1)/2} \\ &\equiv -4Q + 2Q \equiv -2Q \pmod N. \end{aligned} \tag{6.10}$$

Now

$$\begin{aligned} 2\alpha &= 1 + \sqrt{1 - 4Q} = 1 + \alpha - \beta, \\ 2\beta &= 1 - \sqrt{1 - 4Q}. \end{aligned}$$

Raising both sides of these equations to the Nth power and adding and subtracting the results we obtain, using the divisibility of the binomial coefficients by the prime N,

$$\begin{aligned} V_N &\equiv 1 \pmod N, \\ U_N &\equiv (1 - 4Q)^{(N-1)/2} \equiv -1 \pmod N. \end{aligned}$$

From the identity

$$2V_{N+1} = V_N + (1 - 4Q)U_N$$

we see that

$$2V_{N+1} \equiv 1 - (1 - 4Q) \equiv 4Q \pmod{N}.$$

This contradicts (6.10). Hence N is composite and Theorem 3 is proved.

In case only some of the prime factors of $N + 1$ are known, one can hope to use the following counterpart of Theorem 2 to restrict the prime factors of N.

THEOREM 4: *Let q be a prime factor of $N + 1 = q^\alpha m$ and let*

$$V_{(N+1)/2} \equiv 0 \pmod{N}.$$

If

$$V_{2(N+1)/q} - 2Q^{(N+1)/q}$$

is prime to N, then every prime factor of N is of one of the forms $q^\alpha x \pm 1$.

7. SOLUTION OF THE QUADRATIC CONGRUENCE WITH PRIME MODULUS

Without loss of generality one may suppose that the quadratic congruence is transformed into

$$(7.1) \qquad x^2 \equiv a \pmod{p}, \quad (p \text{ a prime}).$$

At a cost proportional to $\log p$ we can evaluate the symbol $\left(\frac{a}{p}\right)$ and tell whether (7.1) has a solution.

Most texts on number theory that discuss the actual solution of the congruence (7.1) remark that (7.1) is only a special case of a quadratic congruence with an arbitrary modulus m. This in turn is capable of solution by Gauss's method of exclusion, an algorithm of order m. In some texts the reader is shown, as a sort of curiosity, the solution

$$x \equiv \pm a^{k+1} \pmod{p}$$

in case $p = 4k + 3$. Far from being a curiosity, this solution is the most practical one since it is obtained at a cost proportional to $\log p$ by the power algorithm.

For $p = 8k + 5$ there is, similarly, the solution

$$x \equiv \pm a^{k+1} \quad \text{or} \quad \tfrac{1}{2}(4a)^{k+1} \pmod{p}$$

according as

$$a^{2k+1} \equiv 1 \quad \text{or} \quad -1 \pmod{p}.$$

For $p = 8k + 1$ the usual statement made is that only the exclusion method is available. On the contrary, a procedure of order $\log p$ follows from the following theorem.

THEOREM 5: *Let p be a prime of the form $4k + 1$ and let h be chosen so that $h^2 - 4a$ is a quadratic nonresidue of p. Let α, β be the roots of*

$$x^2 - hx + a = 0$$

and let

$$V_n = \alpha^n + \beta^n.$$

Then the solutions of

$$x^2 \equiv a \pmod{p}$$

are

$$x \equiv \pm\tfrac{1}{2}V_{(p+1)/2} \pmod{p}.$$

Proof: As in Theorem 3 we introduce

$$U_n = \frac{(\alpha^n - \beta^n)}{(\alpha - \beta)}$$

and note that, because $(\alpha - \beta)^2 = h^2 - 4a$ is a nonresidue of p,

$$U_{p+1} = U_{(p+1)/2}V_{(p+1)/2} \equiv 0 \pmod{p}. \tag{7.2}$$

Also

$$V^2_{(p+1)/2} - (h^2 - 4a)U^2_{(p+1)/2} = 4a^{(p+1)/2} \equiv 4a \pmod{p}. \tag{7.3}$$

By (7.2) either the first or the second term of the left side of (7.3) is divisible by p. The first alternative leads to the conclusion that $h^2 - 4a$ is a quadratic residue of p, contrary to our choice of h. Hence the second alternative must hold, implying the conclusion of the theorem.

In applying the theorem we may follow the same general procedure used in connection with Theorem 3. However, in this more general situation, (6.4) becomes

$$V_n = hV_{n-1} - aV_{n-2}$$

and (6.5) becomes

$$V_{2m} = V_m^2 - 2a^m.$$

Thus, in all cases, the quadratic congruence is a problem of order $\log p$.

8. REPRESENTATION OF PRIMES BY BINARY QUADRATIC FORMS

Let $p = 4k + 1$ be a prime. Then a and b exist such that

$$p = a^2 + b^2, \quad (a \equiv 1 \bmod 4).$$

To find a and b it is sufficient to find a. According to a beautiful theorem of Gauss, a is the absolutely least remainder on division of $\frac{1}{2}\binom{2k}{k}$ by p. This determination is a procedure of order p. Another answer, due to Jacobsthal, is the quadratic character sum

$$a = -\sum_{j=1}^{(p-1)/2} \left(\frac{j}{p}\right)\left(\frac{j^2+1}{p}\right)$$

which is a calculation of order p. Still another method is that of excluding all impossible values of $a < \sqrt{p}$ by a sieve process, a procedure of order $\sqrt{p}$. As a matter of fact, there is a procedure of order $\log p$. This follows from an idea of Hermite.

In fact, by one of the methods of the preceding section, one can solve the congruence

$$x^2 \equiv -1 \pmod{p}, \quad 0 < x < \frac{p}{2}$$

uniquely. Next one applies the Euclidean algorithm to expand the first few terms in the regular continued fraction for x/p. One terminates this process as soon as one reaches a convergent A_n/B_n for which

$$B_n < \sqrt{p} < B_{n+1};$$

then

$$a = \pm B_n \quad \text{or} \quad \pm(xB_n - pA_n),$$

whichever is of the form $4x + 1$; the even number is $\pm b$. To see why this is so, we recall from the theory of continued fractions that

$$\frac{x}{p} - \frac{A_n}{B_n} = \frac{\theta}{B_n B_{n+1}}, \quad |\theta| < 1.$$

Multiplying by pB_n and squaring we have

$$(8.1) \qquad (xB_n - pA_n)^2 = \frac{\theta^2 p^2}{B_{n+1}^2} < p.$$

Adding B_n^2 to both sides gives

$$(xB_n - pA_n)^2 + B_n^2 < p + B_n^2 < 2p.$$

A careful inspection of the left side of this inequality reveals that it is a multiple of p. Hence it is p itself.

It follows that Gauss' coefficient modulo p and Jacobsthal's character sum can be calculated at a cost of order $\log p$.

As for other quadratic forms, the above scheme can be modified to apply to the problem

$$\lambda p = a^2 + Db^2, \quad (D \neq 0).$$

One has only to add DB_n^2, rather than B_n^2, to both sides of (8.1). For $D = 1, 2, 3$, or 7 one always gets $\lambda = 1$. However, for $D = 5$, for example, one gets

$$p = a^2 + 5b^2, \quad p = 20m + 1, 9,$$

or

$$2p = a^2 + 5b^2, \quad p = 20m + 3, 7.$$

Such calculations for large p, beyond existing tables of "quadratic partitions" [1], can be used in evaluating certain functions occurring in cyclotomy such as Jacobi functions, cyclotomic numbers, generalized Gauss sums, other exponential sums and the number of solutions of certain congruences in two variables.

9. MULTIPRECISE ARITHMETIC

Before leaving the subject of algorithms of order $\log n$ it is desirable to make a few comments about multiprecision arithmetic and to correct an impression that the reader may have gathered from the preceding discussion. If a computation involves only $O(\log n)$ steps, it becomes available for very large values of n and it will ultimately find use (by Parkinson's law) for numbers n of many hundreds of words, in fact until the capacity of the machine or the available time is completely filled. This means that we must be prepared to perform the rational operations with an arbitrary number of words for each operand. In the foregoing more or less theoretical discussion, we have tacitly assumed that the cost of an operation, such as multiplication, was independent of the size of n. In fact, the cost is strictly proportional to $(\log n)^2$ for multiplication or division and $\log n$ for addition or subtraction. In actual practice then the theoretical estimates $O(\log n)$ given above must be realistically raised to $O(\log n)^3$ since they involve multiplication or division in most of their $O(\log n)$ steps.

The preparation and development of a library of subroutines for arbitrarily multiprecise operations should be one of the early activities in setting up a number theoretic computing project. It is hardly necessary to emphasize the importance of polishing these subroutines for speed and ease of entering.

A generally useful library would include subroutines for the following operations.

1. Input with decimal to binary conversion.
2. Output with binary to decimal conversion.
3. Addition and subtraction.
4. Multiplication.
5. Division with remainder.
6. Square root with remainder.
7. Greatest common divisor.
8. Chinese remainder algorithm.
9. Jacobi's symbol.
10. Power algorithm $b^n \equiv r \pmod{N}$.

Algorithms of order $\sqrt{n}$

If the cost of an algorithm is to be $O(\sqrt{n})$, then n cannot be very large unless the constant implied by the O symbol is extremely small. Thus we are now dealing, for the most part, with parameters n small enough to fit into one or two machine words.

The first algorithm of order $\sqrt{n}$ that comes to mind is:

10. FACTORIZATION BY TRIAL DIVISION

Nearly every computer laboratory has its own more or less efficient subroutine for factoring an input number n by searching directly for possible prime factors less than $n^{1/2}$. Strictly speaking, the number of primes less than $n^{1/2}$ is more like $2\sqrt{n}/\log n$. However, unless the machine is supplied with a big list of primes the routine has to produce the next likely trial divisor itself and so the total effort is often nearly $O(\sqrt{n})$.

Even when n is so large that it is too expensive to complete the search for possible prime factors of n the routine may be used to find small factors, if any, and to set a lower bound for the smallest unknown factor of n. This bound can be very useful in attempting to factor n by more sophisticated methods. In any case, the routine may be well worth trying since most numbers n have at least one small factor. For example, the probability that a large n has a prime factor less than 10,000 is almost .95.

However, the probability that n has its least prime factor $< 10^6$, given that it exceeds 10,000, is only about $\frac{1}{3}$.

Since we are considering trial divisors small enough to be held by one machine word, ordinary division (as opposed to multiprecise division) will suffice even in the event that n is made up of several machine words. Thus if one sets a limit on the trial divisor, the algorithm becomes one of order $\log n$ for very large n.

Many such routines use an "idiot" method to obtain the next trial divisor d, usually "$d + 2$ replaces d." This has the advantage that all further division can be dispensed with once d exceeds $2n^{1/3}$, using a device suggested by Alway [2]. However, this simple method becomes wasteful as d gets large. For example, between 10^6 and 2×10^6 there are 500,000 odd numbers but only 70,435 primes. Hence in this range one can afford to spend the equivalent of a few division times selecting the next trial divisor. Some routines employ a small sieve to screen out multiples of 2, 3, 5, and 7. This cuts the 500,000 down to 228,571, almost doubling the speed. The extreme effort to eliminate useless divisions by composite trial divisors is to reserve a few thousand words of the memory to store the characteristic binary number

$$.1110110110100110010110 10\cdots$$

whose kth bit is a 0 or 1 according as $2k + 1$ is a prime or not. About 13,888 words of 36 bits each will serve as a list of primes to 10^6. Using intentional overflow or a set of masks one can see to it that all trial divisors are primes. The choice of one or more of the above procedures must be made on economic lines and will depend on the relative speeds of addition and division peculiar to the given machine.

In many cases the number n has its prime factors restricted to one or more arithmetic progressions. For instance, the prime factors of the Mersenne number $2^p - 1$ are all of the form $px + 1$. Since 2 is a quadratic residue of $2^p - 1$, the factors are further restricted to the two progressions $8y \pm 1$. The utilization of such information is, of course, highly desirable even though this complicates the sifting of trial divisors. In fact, there is often a good case for some effort being devoted to the preliminary search for further small quadratic residues of n, the better to restrict its prime factors. From this point of view, the forthright attack on n by consecutive odd trial divisors is naïve indeed. It goès almost without saying that any major attack on n should be preceded by a test for primality costing $O(\log n)$ as explained in Section 7.

The problem of discovering small quadratic residues of a large number n introduces the following algorithm of order $\sqrt{n}$.

11. CONTINUED FRACTION EXPANSION OF $\sqrt{n}$

If we denote the kth complete quotient in the regular continued fraction expansion

$$\sqrt{n} = a_1 + \frac{1}{a_2 +}\frac{1}{a_3 +} \cdots \frac{1}{+ x_k}$$

by

$$x_k = \frac{P_k + \sqrt{n}}{Q_k} = a_k + \frac{1}{x_{k+1}},$$

then it is well known that

$$A_{k-1}^2 - nB_{k-1}^2 = (-1)^{k-1}Q_k,$$

where $A_0 = 1$, $B_0 = 0$, $A_1 = a_1$, $B_1 = 1, \cdots$, are the numerators and denominators of the convergents. It follows that $(-1)^{k-1}Q_k$ is a quadratic residue of n. By factoring many Q's, combining them multiplicatively and eliminating square factors, a collection of small quadratic residues R can be isolated. This procedure can easily be fully automated. Each such R restricts the possible divisors of n by a factor of $\frac{1}{2}$ by assigning these divisors to certain well-tabulated sets of arithmetic progressions. The method is due to Legendre, 1798. The combination of all these restrictions on the trial divisors is a sieve process to be discussed in Section 15. Often the actual factors of n are revealed by the continued fraction itself [3].

Another reason for having in one's subroutine library the continued fraction algorithm for $n^{1/2}$ is that it can be used to solve the basic Pell equations

$$x^2 - ny^2 = \pm 1 \quad \text{or} \quad \pm 4, \tag{11.1}$$

or, more generally,

$$x^2 - ny^2 = N, \quad (|N| < n^{1/2}), \tag{11.2}$$

whenever these equations are possible.

Since x/y is a convergent of $\sqrt{n}$, x and y are often very large and so the programming here will involve our multiprecise arithmetic subroutines. As in the case of the greatest common divisor subroutine, 41 per cent of the partial quotients a_i are equal to unity. In these cases no multiplication is needed.

To solve Equations (11.1) or (11.2) or to show that no solution exists we must be prepared to expand the continued fraction until we reach the half period. Although there is no realistic rigorous estimate for the length L of the half period, there is heuristic reasoning to support the estimate

$$L < .15n^{1/2} \log n.$$

For this reason we have placed this fundamental algorithm under those of order $\sqrt{n}$.

12. REPRESENTATION OF NUMBERS BY BINARY QUADRATIC FORMS

We have already considered the problem of representing a prime p by a binary quadratic form by means of an algorithm of order $\log p$. More generally the representations of m by

$$x^2 + Dy^2 = m, \tag{12.1}$$

where the factorization of m is known, can be found from the representations of its prime factors making use of "conformal multiplication," namely Brahmagupta's identity

$$(x_1^2 + Dy_1^2)(x_2^2 + Dy_2^2) = (x_1x_2 \pm Dy_1y_2)^2 + D(x_1y_2 \pm x_2y_1)^2.$$

The whole procedure has a cost of $O(\log m)$.

The situation changes drastically when the factorization of m is unknown. No longer can we use Fermat's theorem and the power algorithm. We could use Euler's generalization of Fermat's theorem

$$x^{\phi(m)} \equiv 1 \pmod{m},$$

except that $\phi(m)$ is unknown to us. The connection between the representation problem and the factoring problem is especially clear in the case of $D = -1$ ($m \neq 4h + 2$) since

$$x^2 - y^2 = m$$

has the solution

$$x = \frac{p+q}{2}, \quad y = \frac{p-q}{2}$$

for every pair (p, q) of positive integers of like parity whose product is m.

We do not mean to imply that one must first factor m in order to represent m by a binary quadratic form. On the contrary, the theory of representation was developed by Gauss and Legendre in hopes that it would provide new methods of factoring.

Attacking Equation (12.1) directly by Gauss's method of exclusion gives an algorithm of order $\sqrt{m}$. In fact, if $D > 0$ one may search for y. Since $y < (m/D)^{1/2}$, the problem in general increases in cost with the square root of m. One excludes impossible values of y by writing (12.1) as a congruence

$$Dy^2 - m \equiv x^2 \pmod{E}, \tag{12.2}$$

where E is a convenient excluding modulus. One eliminates all y's modulo E for which the left side of (12.2) is not a quadratic residue of E. Choosing about .7 log (m/D) relatively prime values of E and combining the information thus produced gives a sieve procedure powerful enough to eliminate all but a mere handful of y's under $(m/D)^{1/2}$. Of course, one could alternatively search for x. In case $D < 0$, (12.1) is an indefinite form and its solutions (x, y), if any, are infinitely numerous and lie on a hyperbola rather than an ellipse. One seems, at first thought, to lose control over the size of y. On the contrary, if $-D$ is not a square, we again obtain an algorithm of order $\sqrt{m}$. In fact, by a theorem of Chebyshev, if (12.1) has a solution at all there is one with

$$0 < y < \left\{\frac{-(T-1)m}{(2D)}\right\}^{1/2},$$

where (T, U) is the fundamental solution of the Pell equation

$$t^2 - |D|u^2 = 1.$$

For example, if $D = -2$ there is a solution with $y < (m/2)^{1/2}$ or none at all.

If two solutions of (12.1) are found less than the above mentioned limits for y, a factorization of m into two factors results [4]. For example, if one could find a second representation of a Fermat number

$$F_7 = 1 + 2^{128} = 1^2 + (2^{64})^2$$

as the sum of two squares, one could factor F_7.

13. DIFFERENCE OF SQUARES METHOD

The case in which $-D$ is a square is different. We may suppose that $D = -1$ and that m is odd. As we are about to see, the exclusion problem for either x or y in

$$x^2 - y^2 = m \tag{13.1}$$

is now one of order m rather than $m^{1/2}$. Nevertheless we shall discuss it in the present context.

It would be foolhardy to embark on such a problem without first attempting to discover a small prime factor of m. We may, therefore, suppose that m has no prime factor $< w < \sqrt{n}$ so that

$$0 < x - y < w, \qquad \sqrt{N} < x + y < \frac{N}{w}. \tag{13.2}$$

From (13.1) and (13.2) we deduce the following ranges for x and y:

$$0 < y < \frac{(m/w) - w}{2},$$

$$\sqrt{m} < x < \frac{(m/w) + w}{2}.$$

With fixed w it is apparent that the search for either x or y is a problem of order m. However, there are cases in which this method of higher order is actually superior to the preceding quadratic form methods. These are the cases in which it is known that all the prime factors of m are of the form $kz + 1$ for some $k > 2$, as for example in the conclusion of Theorem 2 or whenever m is a primitive factor of $a^k - b^k$ for integers a and b.

Setting

$$x - y = kz_1 + 1, \qquad x + y = kz_2 + 1,$$

we have

$$N + 1 = (x - y)(x + y) + 1 \equiv k(z_1 + z_2) + 2 \equiv 2x \pmod{k^2}.$$

Thus x is restricted to one residue class modulo k^2 or $k^2/2$ according as k is odd or even. In many such cases this restriction is strong enough to make the representation of m by $x^2 - y^2$ cheaper than that by $x^2 + y^2$, for example.

Of course the decision depends also on w. Theoretically, the break even point is where $wk^2 = \frac{1}{2}m^{1/2}$ or $m^{1/2}$ according as k is odd or even. In actual practice much depends on the restrictions imposed by the very small primes and their powers when used as excluding moduli. For details see [5], [6].

More generally, as in Theorem 4 and for primitive divisors of Lucas' function $U_n = (\alpha^n - \beta^n)/(\alpha - \beta)$, one knows that all the prime factors of m are of the form $kz \pm 1$. Similar restrictions to two residue classes modulo k^2 or $k^2/2$ may be found for x or for y according as $N \equiv 1$ or $-1 \pmod{k}$. For details see [7].

Just as the knowledge of a small quadratic residue of m cuts down the search for the factors of m by one-half so also does it affect the search for x in (13.1). See [8] for particulars.

Problems of Order n or More

Of the multitude of problems of order greater than or equal to n, a conspicuous class arises from inversive methods of solving in a wholesale manner a family of problems of lower order. Thus far we have been considering problems as though they were isolated, as indeed many problems necessarily are. In many cases one is interested in solving a problem for a whole range of values of a parameter and there is some indirect mass production approach that can be used to reduce the actual expense per problem below that of the isolated direct attack. It is perhaps a little surprising to find in many cases that such economies, though considerable, are not great enough to reduce appreciably the order of the over-all problem.

To illustrate this phenomenon we may cite the following examples.

Examples

1. Suppose the problem is to make a list of primes $\leq n$. The direct approach would be to examine every integer $x \leq n$ for primality. Even if we threw out a certain proportion of the x's as obviously composite, the whole program would be of order

$$O\left(\int_1^n \log x \, dx\right) = O(n \log n).$$

The indirect approach would be the Sieve of Eratosthenes. Here for each $p \leqq \sqrt{n}$ we go to the trouble of annihilating all of the $[n/p] - 1$ composite multiples of p that are less than n. This adds up to a program whose order is

$$O\left(\sum_{p<\sqrt{n}} \frac{n}{p}\right) = O(n \log \log n),$$

which is only a little improvement over the direct method. Of course, the actual costs for a given n may be very different depending on the constants implied by each of the O-symbols.

2. Let the problem be to find the unique representation $p = a^2 + b^2$ for each prime $p \leqq n$. For the direct approach we would determine (a, b) with a cost of $O(\log p)$, as described in Section 12. The total effort would, therefore, be of order

$$O\left(\sum_{p \leq n} \log p\right) = O(n)$$

by the prime number theorem.

The indirect approach is to form all sums of two squares $\leq n$ and arrange them in increasing order. Those entries which occur only once are then selected and put out. To insure uniqueness of (a, b) we may take a odd and b even, or alternatively, $a < b$. In either case the whole effort will cost an amount proportional to the number of lattice points inside a circle of radius $\sqrt{n}$, that is, $O(n)$. Orderwise the two approaches are, therefore, equivalent even though the indirect approach is many times cheaper.

14. SPECIAL CASES OF LOWER ORDER

Some problems which belong to a general class of high order are, by a stroke of good fortune, actually of much lower order. An example occurs in the determination of the numerical functions generated by the powers of formal power series. Let $f(x)$ be a power series in x with $f(0) = 1$, for simplicity, and let us write

$$f_k(x) = \{f(x)\}^k = \sum_{n=0}^{\infty} c_k(n)x^n.$$

Suppose that $c_k(n)$ is known for $n = O(1)N$ and we wish to determine $c_r(n)$ for $n = O(1)N$ for some given $r \neq k$, neither r nor k being zero.

This may be done recursively by noting that

$$\{f_r(x)\}^k = \{f_k(x)\}^r$$

implies

$$rf_r(x)f_k'(x) = kf_k(x)f_r'(x).$$

Equating coefficients of x^n on both sides gives

$$knc_r(n) = \sum_{s=1}^{n} c_r(n-s)\{(k+r)s - kn\}c_k(s) \tag{14.1}$$

from which $c_r(n)$ can be determined from $c_r(m)$ with $m < n$. Hence to tabulate $c_r(n)$ for $n = O(1)N$, in general, is going to cost

$$O\left(\sum_{n=1}^{N} n\right) = O(N^2).$$

If, however, k has been chosen so that $f_k(x)$ is lacunary, the cost will be very much less. Examples from additive number theory are

$$k = 1, \qquad f_1(x) = 1 + 2\sum_{n=1}^{\infty} x^{n^2},$$

in which we are concerned with number of representations of n as the sum of r squares, and also

$$f(x) = \prod_{n=1}^{\infty} (1 - x^n)$$

with $k = 1$ and $k = 3$, for which we have the well known lacunary series

$$f_1(x) = \sum_{n=-\infty}^{\infty} (-1)^n x^{(3n^2+n)/2},$$

$$f_3(x) = \sum_{n=0}^{\infty} (2n+1)(-1)^n x^{n(n+1)/2},$$

which are theorems of Euler and Jacobi, respectively, in the theory of partitions. In these cases the number of nonzero terms in (14.1) is $O(\sqrt{n})$ so the total cost is only

$$O\left(\sum_{n=1}^{N} \sqrt{n}\right) = O(N^{3/2}).$$

Thus a table of $p(n)$, the number of unrestricted partitions of n, generated by $f_{-1}(x)$ or a table of Ramanujan's function $\tau(n)$, generated by $xf_{24}(x)$ are basically equally expensive. Here we are ignoring the fact that $\tau(n)$ is multiplicative and $p(n)$ is not. In fact it is sufficient to use (23) only for $n-1$, a prime in the case of $\tau(n)$ since

$$\tau(p^\alpha) = \tau(p)\tau(p^{\alpha-1}) - p^{11}\tau(p^{\alpha-2}).$$

However, for moderate values of N, say $N < 30{,}000$, this fact is hardly worth the extra elaboration of the program needed to exploit it.

It is perhaps of some interest to look at the cost of calculating isolated values of $p(n)$ and $\tau(n)$. For the former function we have the Hardy-Ramanujan-Rademacher series [9] which can be safely truncated after only $\sqrt{n}$ terms. Hence we can attach a cost of $O(\sqrt{n})$ to the isolated value of $p(n)$. This is in agreement with the cost per entry of the table of $p(n)$.

For $\tau(n)$ the cost for an isolated value of n is of order n. Perhaps the cheapest formula is the following, for n odd,

$$\tau(n) = n^2\sigma_3(n) + 120 \sum_{k=1}^{(n-1)/2} (2n-3k)(n-3k)\sigma_3(k)\sigma_3(n-k),$$

which is due to van der Pol [10]. Here $\sigma_3(n)$ denotes the sum of the cubes of the divisors of n.

15. SIEVE PROCEDURES

We have already referred to exclusion methods in Section 12 in connection with the representation of a given number by a binary quadratic form. The method of exclusion is widely applicable to Diophantine equations and many other questions and constitutes a powerful tool. There are various techniques for automating the actual exclusion and these come under the heading of Sieve Procedures and are part of the set inclusion problem of combinatorics [11]. To use the logical facilities of the arithmetic unit, one constructs a certain number of "wheels" in the memory, one for each excluding modulus m_i. Each wheel consists of m_i binary bits. To start with, the kth bit of wheel m_i is a 0 or 1 according as k is or is not acceptable mod m_i. Each wheel rotates by cyclic permutation of period m_i. The amount of rotation per step may be only one bit in the simpler programs or it may be one or more words in the more elaborate programs. The object is to find the first number whose corresponding wheel position is a 0 for each wheel.

In the simpler programs this can be done, for example, by a left shift of 1 and inspection for overflow. In the more elaborate programs one or more words full of bits can be processed at each step by "OR-ing" the wheels together, see [12] for details. The most elaborate sieve program known to the writer has been written by J. D. Brillhart for the IBM 7094. It uses a very few quite large wheels and most of the available storage space and excludes almost all numbers at the rate of 150,000 per second. Correspondingly greater speeds could be achieved by a machine with more bits per word.

16. APPROXIMATE OPERATIONS

Thus far, with one exception, we have discussed algorithms and operations carried out with rational integers. The one exception was the bare mention of the Hardy-Ramanujan-Rademacher formula for the partition function

$$p(n) = (24n - 1)^{-1}\sqrt{12} \sum_{|k| \leq \sqrt{n}} A_{|k|}(n)|k|^{-1/2}\left(1 - \frac{k}{\mu}\right) \exp\left(\frac{\mu}{k}\right),$$

where

$$\mu = \frac{\pi}{6}(24n - 1)^{1/2}$$

and $A_k(n)$ is a complicated sum of $24k$th roots of unity. This is a good example of the use of approximations to irrational numbers to determine exact integers.

Other more routine instances are numerous and include the exact determination of number of zeros of the Riemann zeta-function in a finite critical strip and the evaluation of algebraically complicated symmetric functions of a set of conjugate algebraic integers.

When the number theorist thus ventures into the realm of the computer-simulated real and complex number systems, he may well find his requirements, both as to accuracy and to magnitude, may exceed those provided for the differential equations man by the system's subroutines. In such cases it may be useful to prepare one's own set of floating point subroutines using a full machine word for the characteristic and two or three machine words for the mantissa. This is especially advisable when seeking sufficiently accurate roots of polynomial equations with integer coefficients.

17. ALGEBRAIC NUMBER THEORY

In dealing with irrational algebraic numbers one need not necessarily resort to the use of approximations. Instead, the elements of a ring of algebraic integers or the ideals of such a ring can be represented exactly and conveniently by a vector of rational integers in terms of a basis. Similarly the elements of an algebraic number field can be rationalized and represented by a pair (V, d) where V is a vector of rational integers and d is the rational integral denominator.

Operations in terms of such representations including addition, multiplication, norming, greatest common divisor, if possible, congruential operations, exponentiation, etc., all involve manipulations with vectors and are best handled through subroutines.

In such problems it is usual to assign a machine word to each component of a vector, with perhaps a further word to hold the dimension of the vector. In some cases, however, the components may grow in size, as for example in the exact continued fraction expansion of an algebraic number of degree greater than two. Recourse must then be had to multiprecise arithmetic subroutines, in which two or more words are needed for some of the components, with even possibly a "push down" listing of each vector. In other cases, when there is need for the utmost speed, vectors with a modest number of small components may be packed one vector to a machine word. In this packed format a vector can be sent from place to place in minimum time. Certain simple operations on the separate components can be performed in parallel by the machine's logical repertory. Other less simple operations can be performed in series by "dribbling" and "masking" components in the registers of the arithmetic unit. For details see [11]. Needless to say, such operations are impossible to describe in a language, like ALGOL, that fails to take into account the word structure of the machine.

In some cases, instead of using bases and vectors, one can recast conditions and statements involving irrational algebraic integers or ideals so as to obtain practical expressions involving only rational integers. This can be particularly desirable in dealing with fields of high degree as, for example, the cyclotomic fields of the Fermat problem. For an instance of this procedure see [13].

18. REPRESENTATION OF INFINITE CLASSES

The same vectors discussed above can be used to represent a possibly large finite number of infinite sets. This fact is useful in cases where infinite sets must be processed systematically as in some kinds of theorem proving. To illustrate, suppose that S is the set of all positive integers having no prime factor greater than 11. Then every member of S is a product of a cube belonging to S and a number of the form

$$m = 2^{\alpha_1}3^{\alpha_2}5^{\alpha_3}7^{\alpha_4}11^{\alpha_5}, \quad (0 \leqq \alpha_i \leqq 2).$$

We may represent m by the vector

$$[\alpha_1, \alpha_2, \alpha_3, \alpha_4, \alpha_5]. \tag{18.1}$$

There are precisely $243 = 3^5$ such vectors, each one representing an infinite subset of S. Again, corresponding to any vector

$$[\beta_1, \beta_2, \beta_3, \beta_4, \beta_5], \quad (0 \leqq \beta_i \leqq 2), \tag{18.2}$$

we may consider the infinite set S' of those primes $p > 13$ of the form $6x + 1$ whose indices of the first five primes satisfy

$$\text{Ind } 2 \equiv \beta_1 \pmod 3,$$

$$\text{Ind } 3 \equiv \beta_2 \pmod 3,$$

$$\text{Ind } 5 \equiv \beta_3 \pmod 3,$$

$$\text{Ind } 7 \equiv \beta_4 \pmod{3},$$
$$\text{Ind } 11 \equiv \beta_5 \pmod{3}.$$

The inner product of the vectors (18.1) and (18.2) taken modulo 3 indicates the cubic character of every member of S with respect to every prime modulus of S'. For further examples of this technique see [14], [15].

BIBLIOGRAPHY

1. Cunningham, A. J. C., *Quadratic Partitions*. London, 1907; *Quadratic and Linear Tables*. London, 1927.
2. Alway, G. G., "A method of factorisation using a high-speed computer," *Math. Comp.*, 6 (1952), 59–60.
3. Lehmer, D. H., and Powers, R. E., "On factoring large numbers," *Amer. Math. Soc. Bull.*, 37 (1931), 770–776.
4. Mathews, G. B., *Theory of Numbers*. Cambridge, 1892, 266–267.
5. Kraitchik, M., *Théorie des Nombres*, Volume 1. Paris, 1922.
6. ———, *Recherches sur la Théorie des Nombres*, Volume 2. Paris, 1929.
7. Lehmer, D. H., "On the factorization of Lucas' functions," *Tôhoku Math. J.*, 34 (1931), 1–7.
8. Kraitchik, M., *Théorie des Nombres*, Volume 2. Paris, 1926.
9. Lehmer, D. H., "On the series for the partition function," *Amer. Math. Soc. Trans.*, 43 (1938), 271–295.
10. Vander Pol, B., "On a nonlinear partial differential equation satisfied by the logarithm of the Jacobian theta-functions, with arithmetical applications," *Nederl. Akad. Wetensch. Proc., Series A*, 54 (1951), 261–284.
11. Lehmer, D. H., *Applied Combinatorial Mathematics*, Chapter 1. New York, 1964.
12. ———, "The sieve problem for all-purpose computers," *Math. Comp.*, 7 (1953), 6–14.
13. Lehmer, D. H., Lehmer, Emma, and Vandiver, H. S., "An application of high-speed computing to Fermat's last theorem," *Nat. Acad. Sci., Washington, Proc.*, 40 (1954), 25–33.

14. Lehmer, D. H., Lehmer, Emma, Mills, W. H., and Selfridge, J. L., "Machine proof of a theorem on cubic residues," *Math. Comp.*, 16 (1962), 407–415.

15. Brillhart, J. D., Lehmer, D. H., and Lehmer, Emma, "Bounds for pairs of consecutive seventh and higher power residues," *Math. Comp.*, 18 (1964), 397–407.

ASYMPTOTIC DISTRIBUTION OF BEURLING'S GENERALIZED PRIME NUMBERS

Paul T. Bateman
and
Harold G. Diamond

1. INTRODUCTION

During the present century a considerable portion of the work done on the distribution of primes has been concerned with investigating the "stability" of the prime number theorem and other theorems on the distribution of primes. Many mathematicians have tried to determine how general these theorems are and what is really needed to prove them.

The prime number theorem (P.N.T.) states that if $\pi(x)$ is the number of primes not exceeding the positive real number x, then

$$\lim_{x\to\infty} \frac{\pi(x)}{x/\log x} = 1. \tag{1.1}$$

Thus if x is large, the fraction of positive integers up to x which are primes is about $1/\log x$. For example, about one per cent of the positive integers less than e^{100} are primes.

The first crude observation to be made is that the additive structure of the positive integers is not particularly relevant to the distribution of primes. As an example of this phenomenon, we consider the multiplicative semigroup S of positive integers generated by the primes greater than 4. Except for the loss of the primes 2 and 3, the generators of S are the same as the generators of the semigroup of all positive integers, and so the distribution of the generators of S is substantially the same as that of the ordinary primes. However, S has no additive structure, since it consists of the positive integers relatively prime to 6, i.e., congruent to 1 or 5 modulo 6. If $\phi(x; 6)$ is the number of elements of S not exceeding x, then it is easy to verify that

$$\phi(x; 6) = \left[\frac{x+1}{6}\right] + \left[\frac{x+5}{6}\right], \qquad \left|\phi(x; 6) - \frac{x}{3}\right| \leqq \frac{2}{3}.$$

(Throughout this article $[y]$ denotes the greatest integer not exceeding y.) Thus, although S differs from the set of all positive integers in having no additive structure, it does retain the property that the number of its elements not exceeding x is very well approximated by a constant times x. More generally, if $\phi(x; k)$ denotes the number of positive integers not exceeding x which are relatively prime to the fixed positive integer k, then

$$\phi(x; k) = \sum_{1 \leqq h \leqq k, (h,k)=1} \left[\frac{x-h}{k} + 1\right], \qquad \left|\phi(x; k) - \frac{\phi(k)}{k} x\right| < \phi(k),$$

where $\phi(k)$ is the number of positive integers not exceeding k which are relatively prime to k.

As a less trivial example we consider the sequence of positive integers $(p_1, p_2, \cdots)$ we get by listing in nondecreasing order the norms of the distinct prime ideals of the ring of integers of some particular algebraic number field K. Then the norms of the integral ideals are the values of the product $\prod_{j=1}^{\infty} p_j^{\nu_j}$, where $(\nu_1, \nu_2, \cdots)$ is an arbitrary sequence of nonnegative integers all but a finite number of which are zero. We know in a general way how these norms of integral ideals are distributed. In fact, let $I(x)$ denote the number of integral ideals with norm not exceeding x, that is, let $I(x)$ be the number of sequences $(\nu_1, \nu_2, \cdots)$ of nonnegative integers such that $\prod_{j=1}^{\infty} p_j^{\nu_j} \leqq x$. In terms of our notation, Weber (cf. [27] and [28]) proved that

$$I(x) = Ax + O(x^{\theta}), \tag{1.2}$$

where A and θ are constants depending on the field with $A > 0$ and $\theta < 1$. (Cf. Appendix 1 for the definitions of O and o.) The assertion (1.2) can be considered as an analogue of the assertion $[x] = x + O(1)$. If $\pi(x)$ now denotes the number of prime ideals with norms not exceeding x, that is, $\pi(x) = \sum_{p_i \leqq x} 1$, then the relation (1.1) still holds, as Landau proved in 1903 [13].

An examination of Landau's proof shows that (1.2) is the only essential fact about algebraic number fields which is used. This suggests that we consider the following general situation. Suppose we have a sequence P of positive real numbers $(p_1, p_2, \cdots)$ such that

$$1 < p_1 \leqq p_2 \leqq \cdots, \quad p_j \to \infty.$$

We shall call such a sequence P a *generalized prime number system*, or briefly, a system of g-primes. The reason for this terminology will be made clear presently. We shall denote by $\pi_P(x)$ the number of positive integers j such that $p_j \leq x$. When the context is clear, we shall suppress the P and write simply $\pi(x)$. The mild assumptions made above show that for each real number x there are only finitely many sequences $(\nu_1, \nu_2, \cdots)$ of nonnegative integers for which

$$\prod_{j=1}^{\infty} p_j^{\nu_j} \leqq x. \tag{1.3}$$

[This inequality implies that only finitely many terms in the sequence $(\nu_1, \nu_2, \cdots)$ are different from zero. In particular, if $p_J > x$, then $\nu_j = 0$ for all $j \geqq J$.] We define a function $I = I_P$ on $[1, \infty)$ by taking $I_P(x)$ to be the number of sequences $(\nu_1, \nu_2, \cdots)$ satisfying (1.3).

Set $n_0 = 1$ and let $(n_0, n_1(=p_1), n_2, n_3, \cdots)$ be the nondecreasing sequence of real numbers formed by the values of $\prod_{j=1}^{\infty} p_j^{\nu_j}$, where each ν_j is allowed to range over all nonnegative integers. The sequence $(n_0, n_1, n_2, \cdots)$ will be called the *generalized integers of the system P*, or briefly, the g-integers associated with P, for these numbers form the multiplicative semigroup generated by P, in the same way as the ordinary positive integers are generated by the ordinary primes.

We have not required the numbers n_i to be all distinct. Indeed, if

$$n_{k-1} < y = n_k = n_{k+1} = \cdots = n_{k+m-1} < n_{k+m},$$

then we shall call y a g-integer of multiplicity m. In other words, there are m distinct sequences $(\nu_1, \nu_2, \cdots)$ such that $\prod_{j=1}^{\infty} p_j^{\nu_j} = y$. Also, we then have

$$I_P(y) - \lim_{x \uparrow y} I_P(x) = m.$$

The definitions of g-primes and g-integers are due to Beurling [3].

For P a g-prime system, we are interested in the relationship between the two sequences $(p_1, p_2, \cdots)$ and $(n_0, n_1, n_2, \cdots)$ or between the functions π_P and I_P.

In terms of this notation (which we shall use throughout this article), Landau (*op. cit.*) proved (essentially) the following theorem.

If P is a g-prime system such that

$$I_P(x) = Ax + O(x^\theta) \quad (x \geqq 1), \tag{1.4}$$

where $A > 0$ *and* $\theta < 1$, *then*

$$\pi_P(x) = \frac{x}{\log x} + o\left(\frac{x}{\log x}\right), \tag{1.5}$$

i.e., the P.N.T. holds for P.

Actually, Landau's original method of proof gives the more precise estimate

$$\pi_P(x) = \int_2^x \frac{du}{\log u} + O(x \exp\{-c \log^\lambda x\})$$

for any fixed $\lambda \in (0, \frac{1}{8})$ and some $c > 0$ (cf. [15]). A subsequent method developed by Landau [16] raised λ to $\frac{1}{2}$. (l'Hôpital's rule shows that the ratio of $x/\log x$ to $\int_2^x (\log u)^{-1}\, du$ tends to 1 as $x \to \infty$. Thus we may use either expression as an approximation to $\pi(x)$ in the statement of the prime number theorem. For ordinary primes {and many other cases} the integral provides a closer approximation to $\pi(x)$ than does $x/\log x$ {cf., e.g., [10] or [23]}.)

Landau's results show that the assumption (1.4) gives much more than (1.5) alone. It is natural to ask to what extent we may weaken the assumptions on the function I_P and still be able to prove the P.N.T. This problem was raised by Beurling in 1937 [3]. Beurling showed that *if*

$$I_P(x) = Ax + O(x \log^{-\gamma} x), \tag{1.6}$$

where A is a positive constant and $\gamma > \frac{3}{2}$, *then* (1.5) *holds.*

If $\gamma = \frac{3}{2}$ in assumption (1.6), the P.N.T. need not hold, as Beurling showed by a continuous analogue of a generalized prime number system. Diamond [6] has recently shown how to alter Beurling's example to give a set of discrete primes by taking $P = (p_1, p_2, \cdots)$, where p_j is the (unique) real number x satisfying the equation

$$\int_1^x (\log t)^{-1}(1 - \cos\{\log t\})\, dt = j.$$

Thus, Beurling's theorem tells us that the P.N.T. is "reasonably stable" under perturbation of the integers, while the above examples show that there is a limit to its stability.

This article will be devoted primarily to giving a proof of the positive portion of Beurling's theorem. Our method of proof follows the general outline of Beurling's original paper [3]. In many of its details, however, our argument is closer to Landau's proof of the P.N.T. [14]. The reader interested in seeing a proof of Beurling's theorem with a minimal tour through the countryside can omit Sections 3, 4, and 6.

Many theorems expressed in terms of the multiplicative structure of the positive integers are even more stable than the prime number theorem. In a way this says that such theorems are more superficial. As examples, we shall obtain generalizations of the theorem that $\pi(x)/[x] \to 0$ and of Rényi's theorem [24] that *the positive integers n with* $\Omega(n) - \omega(n) = k$ *have a density* ρ_k *given by the power series*

$$\sum_{k=0}^{\infty} \rho_k z^k = \prod_p \left\{1 + \frac{z-1}{p(p-z)}\right\} \quad (|z| < 2),$$

where the product is taken over all the primes. Here $\omega(n)$ is the number of distinct prime factors of n, and $\Omega(n)$ is the total number of prime factors of n counting repetitions. The assumption needed for generalizing both these theorems is that $I_P(x) \sim xL(x)$, where L is a slowly varying function, i.e., a function on $[1, \infty)$ such that $L(cx)/L(x) \to 1$ as $x \to \infty$ for any fixed positive c. Our generalization of Rényi's theorem will be obtained by the method used by Kac [11] in his proof of Rényi's theorem.

The very strong assumption (1.4) enables one to carry over any reasonable asymptotic formula based on the multiplicative structure of the positive integers without any substantial change in the proof. Theorems such as Beurling's theorem and our generalization of Rényi's theorem are challenging because they involve skating on much thinner ice.

We have considered the stability of the prime number theorem under perturbations of the formula for the number of integers not exceeding x. A totally different way of considering its stability is to see what happens if we make perturbations in the sieve of Eratosthenes. A remarkable stability still shows up (cf. [32] and the references cited therein), but we shall not enter into that subject.

We have assumed some standard results from analysis, though we have given less familiar definitions and theorems in the appendix. All the analysis that we need may be found, for instance, in Apostol [2]. We have tried to make our number theoretic discussion self contained. However, our arguments will certainly be more meaningful to the person acquainted with the basic facts about the ordinary prime numbers. For this background, we refer the reader to such standard books as Hardy and Wright [8] or the first two chapters of the monograph of Ingham [10].

2. ZETA FUNCTIONS OF GENERALIZED PRIME SYSTEMS

If we are given a generalized prime number system P, then formally we have

$$\sum_{i=0}^{\infty} n_i^{-s} = \prod_{j=1}^{\infty} \sum_{\nu_j=0}^{\infty} p_j^{-\nu_j s} = \prod_{j=1}^{\infty} (1 - p_j^{-s})^{-1}, \tag{2.1}$$

where s is a complex variable and x^{-s} is defined for $x \geqq 1$ to be $\exp(-s \log x)$, where $\log x$ is real. We shall be concerned only with generalized prime number systems in which the formal identity (2.1) makes sense as an equality between analytic functions, for at least some values of the variable s.

LEMMA 2A: *For any generalized prime number system P, the series $\sum_{i=0}^{\infty} n_i^{-\sigma}$, the series $\sum_{j=1}^{\infty} p_j^{-\sigma}$, and the product $\prod_{j=1}^{\infty} (1 - p_j^{-\sigma})^{-1}$ converge for exactly the same positive real values of σ. If they all converge for $\sigma = \sigma_0$, then $\sum_{i=0}^{\infty} n_i^{-s}$, $\sum_{j=1}^{\infty} p_j^{-s}$, and $\prod_{j=1}^{\infty} (1 - p_j^{-s})^{-1}$ converge uniformly on $\{s\colon \operatorname{Re} s \geqq \sigma_0\}$, and on this set* (2.1) *holds.*

Proof: The first assertion is a consequence of the following simple inequalities:

$$\begin{aligned} \sum_{p_j < p_N} p_j^{-\sigma} &\leqq \sum_{n_i < p_N} n_i^{-\sigma} \\ &\leqq \sum_{\nu_1=0}^{\infty} \cdots \sum_{\nu_N=0}^{\infty} (p_1^{\nu_1} \cdots p_N^{\nu_N})^{-\sigma} = \prod_{j=1}^{N} (1 - p_j^{-\sigma})^{-1} \end{aligned}$$

and

$$\begin{aligned} \prod_{j=1}^{N} (1 - p_j^{-\sigma})^{-1} &= \prod_{j=1}^{N} \left(1 + \frac{p_j^{-\sigma}}{1 - p_j^{-\sigma}}\right) \\ &\leqq \prod_{j=1}^{N} \exp \frac{p_j^{-\sigma}}{1 - p_j^{-\sigma}} \leqq \exp\left\{\frac{1}{1 - p_1^{-\sigma}} \sum_{j=1}^{N} p_j^{-\sigma}\right\}. \end{aligned}$$

(The inequality preceding the first "exp" uses the well known relation $1 + x \leqq e^x$.) Note that the above inequalities are valid even if p_N has multiplicity greater than 1.

We shall prove part of the second assertion of the lemma by using the Cauchy convergence principle. If M and N are integers and $M \leqq N$, we have

$$\left| \sum_{j=M}^{N} n_j^{-s} \right| \leqq \sum_{j=M}^{N} n_j^{-\sigma_0} \quad (\text{Re } s \geqq \sigma_0),$$

and since the right side of this inequality tends to zero as $M, N \to \infty$, $\sum n_i^{-s}$ converges uniformly on $\{s\colon \text{Re } s \geqq \sigma_0\}$. Similar remarks apply to $\sum p_j^{-s}$.

To complete the proof, we shall derive a relation connecting partial products and partial sums. Since

$$\prod_{j=1}^{N} (1 - p_j^{-s})^{-1} = \sum_{\nu_1=0}^{\infty} \cdots \sum_{\nu_N=0}^{\infty} (p_1^{\nu_1} \cdots p_N^{\nu_N})^{-s} \quad (\text{Re } s > 0),$$

we have

$$\prod_{j=1}^{N} (1 - p_j^{-s})^{-1} = \sum{}^{*} n_i^{-s},$$

where * indicates that the summation extends over precisely those integers i for which $n_i = p_1^{\nu_1} p_2^{\nu_2} \cdots$ with $\nu_j = 0$ for $j > N$. Also,

$$\sum{}^{*} n_i^{-s} = \sum_{n_i < p_N} n_i^{-s} + \sum_{n_i \geqq p_N}{}^{*} n_i^{-s}.$$

From the last two equalities we deduce an important relation between sums and products:

$$\left| \prod_{j=1}^{N} (1 - p_j^{-s})^{-1} - \sum_{n_i < p_N} n_i^{-s} \right| \leqq \sum_{n_i \geqq p_N}{}^{*} n_i^{-\sigma_0} \quad (\text{Re } s \geqq \sigma_0). \tag{2.2}$$

Now $\sum n_i^{-s}$ converges uniformly on $\text{Re } s \geqq \sigma_0$, and the right side of (2.2) tends to zero as $N \to \infty$. Thus $\prod (1 - p_j^{-s})^{-1}$ converges uniformly for $\text{Re } s \geqq \sigma_0$, and the second assertion of the lemma is proved.

The equality (2.1) now follows immediately from (2.2). ▌

If P is a generalized prime number system, the infimum α of those real values σ for which $\sum n_i^{-\sigma}$, $\sum p_j^{-\sigma}$, and $\prod (1 - p_j^{-\sigma})^{-1}$

converge is called the *abscissa of convergence of P*. Clearly $\alpha \geqq 0$, since we have divergence at $s = 0$. If $\alpha > 0$, we can have either convergence or divergence for $\sigma = \alpha$.

The abscissa of convergence of P is related to the rate of growth of π_P and I_P, as we now show.

LEMMA 2B: *Suppose P is a g-prime system with abscissa of convergence $\alpha \geq 0$. Let $f_\sigma: x \mapsto I_P(x)x^{-\sigma}$ and $g_\sigma: x \mapsto \pi_P(x)x^{-\sigma}$. If $\sigma > \alpha$, f_σ and g_σ are bounded functions on $[1, \infty)$. If $\sigma < \alpha$, f_σ and g_σ are unbounded functions on $[1, \infty)$.*

Proof: If $\sigma > \alpha$, we have

$$\pi(x)x^{-\sigma} \leqq I(x)x^{-\sigma} \leqq \sum_{n_i \leqq x} n_i^{-\sigma} \leqq \sum_{i=0}^{\infty} n_i^{-\sigma} < \infty.$$

Thus f_σ and g_σ are bounded on $[1, \infty)$.

To prove the second statement, write $\sum p_j^{-\sigma}$ as a Stieltjes integral (cf. Appendix 2) and integrate by parts, obtaining

$$\sum_{p_j \leqq x} p_j^{-\sigma} = \int_1^x t^{-\sigma}\, d\pi(t) = \pi(x)x^{-\sigma} + \sigma \int_1^x \pi(t)t^{-\sigma-1}\, dt.$$

If g_{σ_1} is bounded on $[1, \infty)$ and if $\sigma > \sigma_1$, the right hand member of the preceding formula is bounded independently of x. Thus $\sum p_j^{-\sigma}$ converges, and so $\sigma_1 \geqq \alpha$. By contraposition, if $\sigma < \alpha$, g_σ is unbounded, and *a fortiori* f_σ is unbounded. ∎

If P is a g-prime system with abscissa of convergence α, we define the *zeta function of P* by the formula

$$\zeta_P(s) = \sum_{i=0}^{\infty} n_i^{-s} \quad (\operatorname{Re} s > \alpha). \tag{2.3}$$

We shall freely drop the P when the context is clear. In view of Lemma 2A, we have

$$\zeta_P(s) = \prod_{j=1}^{\infty} (1 - p_j^{-s})^{-1} \quad (\operatorname{Re} s > \alpha).$$

This function is a generalization of the Riemann zeta function and is of fundamental importance in our discussion. We list below some of the important properties of ζ_P.

LEMMA 2C: *Suppose P is a g-prime system with abscissa of con-*

vergence α. Then the associated zeta function $\zeta = \zeta_P$ does not vanish on $\{s: \operatorname{Re} s > \alpha\}$, and on this set the following equations hold:

$$\zeta(s) = s \int_1^\infty x^{-s-1} I(x)\, dx \tag{2.4}$$

$$\zeta(s) = \exp \sum_{j=1}^\infty \sum_{k=1}^\infty \frac{1}{k} p_j^{-ks} \tag{2.5}$$

$$\frac{\zeta'}{\zeta}(s) = - \sum_{j=1}^\infty \sum_{k=1}^\infty (\log p_j) p_j^{-ks}. \tag{2.6}$$

Proof: For $\sigma = \operatorname{Re} s > \alpha$, we have

$$|\zeta(s)| = \left| \prod_{j=1}^\infty (1 - p_j^{-s})^{-1} \right| \geqq \prod_{j=1}^\infty (1 + p_j^{-\sigma})^{-1}$$
$$\geqq \prod_{j=1}^\infty \exp(-p_j^{-\sigma}) = \exp\left(- \sum_{j=1}^\infty p_j^{-\sigma} \right) > 0.$$

Again, we have used the simple inequality $1 + x \leqq e^x$.

Equation (2.4) is established by integrating by parts the Stieltjes integral

$$\zeta(s) = \int_{1-}^\infty x^{-s}\, dI(x) = \lim_{X \to \infty} \left\{ I(X) X^{-s} + s \int_1^X I(x) x^{-s-1} dx \right\}$$

and noting that $I(X)X^{-s} = o(1)$ for $\operatorname{Re} s > \alpha$.

To establish Equation (2.5), we set

$$\zeta(s, N) = \prod_{j=1}^N (1 - p_j^{-s})^{-1} = \exp \sum_{j=1}^N \{-\log(1 - p_j^{-s})\}.$$

Using the power series for the principal branch of $\log(1 + z)$ for $|z| < 1$, we get

$$\zeta(s, N) = \exp \sum_{j=1}^N \sum_{k=1}^\infty \frac{p_j^{-ks}}{k}.$$

We proved in Lemma 2A that for $\operatorname{Re} s > \alpha$ the function $\zeta(s,N)$ converges to ζ as $N \to \infty$. On the set $\{s: \operatorname{Re} s \geqq \alpha + \epsilon\}$ the series

$$\sum_j \sum_k \left| \frac{p_j^{-ks}}{k} \right|$$

is majorized by the convergent numerical series

$$\sum_{j=1}^\infty \sum_{k=1}^\infty p_j^{-k(\alpha+\epsilon)} = \sum_{j=1}^\infty \frac{p_j^{-\alpha-\epsilon}}{1 - p_j^{-\alpha-\epsilon}} \leqq (1 - p_1^{-\alpha-\epsilon})^{-1} \sum_{j=1}^\infty p_j^{-\alpha-\epsilon}.$$

Thus $\Sigma\ \Sigma\ p_j^{-ks}/k$ converges absolutely and uniformly for $\operatorname{Re} s \geqq \alpha + \epsilon$, and formula (2.5) follows from the continuity of the exponential function.

We prove Equation (2.6) by differentiating (2.5) term by term on the right-hand side. This step is justified by the uniform convergence of the differentiated series on the set $\operatorname{Re} s \geqq \alpha + \epsilon$ for every $\epsilon > 0$ (cf. Appendix 3). ▮

We introduce here a new function $\psi = \psi_P$ defined on the positive reals by

$$\psi(x) = \sum_{p_j^k \leqq x} \log p_j.$$

Here the summation is over pairs of positive integers (j, k) satisfying the condition $p_j^k \leqq x$. This function is a generalization of Chebyshev's psi function of classical prime number theory. We shall frequently use ψ for the reason that formulas involving it are simpler than corresponding formulas involving π. We collect some of the important properties of ψ in the following lemma.

LEMMA 2D: *Let P be a g-prime system with abscissa of convergence α and let ζ and ψ be as defined above. Then*

$$\psi(x) = \sum_{p_j \leqq x} \left[\frac{\log x}{\log p_j}\right] \log p_j, \tag{2.7}$$

$$-\frac{\zeta'(s)}{\zeta(s)} = \int_1^\infty x^{-s}\, d\psi(x) \quad (\operatorname{Re} s > \alpha), \tag{2.8}$$

$$-\frac{\zeta'(s)}{\zeta(s)} = s\int_1^\infty x^{-s-1}\psi(x)\, dx \quad (\operatorname{Re} s > \alpha). \tag{2.9}$$

Proof: For given $p_j \in P$ there are precisely $\left[\frac{\log x}{\log p_j}\right]$ integers k which satisfy $p_j^k \leqq x$. With this observation, (2.7) follows from the definition of ψ.

The representation (2.8) follows from (2.6), the definition of the Riemann-Stieltjes integral, and the observation that ψ is constant except for jumps at points expressible as p_j^k, where $p_j \in P$ and k is a positive integer.

We need a simple estimate for ψ in terms of π to establish (2.9):

(2.10) $$\psi(x) = \sum_{p_j \leqq x} \left[\frac{\log x}{\log p_j}\right] \log p_j \leqq \sum_{p_j \leqq x} \frac{\log x}{\log p_j} \log p_j = \pi(x) \log x.$$

Now integrating (2.8) by parts, we have

$$-\frac{\zeta'(s)}{\zeta(s)} = \lim_{X\to\infty} \{X^{-s}\psi(X) + s\int_1^X x^{-s-1}\psi(x)\, dx\} \quad (\text{Re } s > \alpha).$$

By (2.10) and Lemma 2B, we have for each number $\epsilon > 0$

$$x^{-(\alpha+\epsilon)}\psi(x) \leqq x^{-(\alpha+\epsilon)}\pi(x) \log x = O(\log x).$$

If Re $s = \alpha + 2\epsilon$ ($\epsilon > 0$), we have

$$\lim_{X\to\infty} |X^{-s}\psi(X)| = \lim_{X\to\infty} X^{-\alpha-\epsilon}\psi(X) \cdot X^{-\epsilon} = 0.$$

We may let $X \to \infty$ in the above expression for $-\zeta'/\zeta$ and thus obtain (2.9). ∎

The prime number theorem is equivalent to the condition $\psi(x) = x + o(x)$, as we now show.

LEMMA 2E: *Let the functions π and ψ be as defined above. Then*

$$\psi(x) = O(x) \text{ iff } \pi(x) = O\left(\frac{x}{\log x}\right).$$

If either of the expressions $x^{-1}\psi(x)$ or $x^{-1}\pi(x)$ log x has a limit as $x \to \infty$, then so does the other, and the limits are equal. In particular, the prime number theorem is true iff $\psi(x) = x + o(x)$.

Proof: We have already shown

(2.10′) $$\psi(x) \leqq \pi(x) \log x.$$

We establish an inequality in the opposite direction. We have

$$\pi(x) = \sum_{p_j \leqq x} 1 \leqq \sum_{p_j^k \leqq x} \frac{\log p_j}{\log p_j^k} = \int_1^x \frac{d\psi(t)}{\log t}.$$

Integrating the last expression by parts, we obtain

(2.11) $$\pi(x) \leqq \frac{\psi(x)}{\log x} + \int_{p_1}^x \frac{\psi(t)}{t \log^2 t}\, dt.$$

Now

$$\int_{p_1}^x \frac{\psi(t)}{t \log^2 t}\, dt \leqq \left\{\max_{1 \leqq t \leqq x} \frac{\psi(t)}{t}\right\} \int_{p_1}^x \frac{dt}{\log^2 t},$$

and for $x \geqq p_1$ we have

$$\int_{p_1}^{x} \frac{dt}{\log^2 t} = \left(\int_{p_1}^{\sqrt{p_1 x}} + \int_{\sqrt{p_1 x}}^{x}\right) \frac{dt}{\log^2 t} \leqq Ax \log^{-2} x,$$

where A is a number which depends only on p_1. Thus,

$$\pi(x) \log x \leqq \psi(x) + \left\{\max_{1 \leqq t \leqq x} \frac{\psi(t)}{t}\right\} \frac{Ax}{\log x}. \tag{2.11'}$$

Now (2.10′) and (2.11′) imply the equivalence of the propositions $\psi(x) = O(x)$ and $\pi(x) = O(x/\log x)$. Moreover, if one of the expressions $x^{-1}\psi(x)$, $x^{-1}\pi(x) \log x$ has a limit as $x \to \infty$, then it is certainly bounded. By using (2.10′), if necessary, we have $\psi(x) \leqq Bx$, for some positive number B. Using (2.10′) and (2.11′) again, we can then show even more than was claimed in the second part of the lemma, namely

$$0 \leqq \frac{\pi(x) \log x}{x} - \frac{\psi(x)}{x} \leqq \frac{AB}{\log x} \quad (x > 1). \blacksquare$$

3. GROWTH OF THE PRIME COUNTING FUNCTION

We give some simple estimates of π_P before making further assumptions on our generalized number system. At the same time we illustrate some of the ways of extracting prime number information from the function ζ_P. Needless to say, we cannot expect anything but trivial theorems with the present paucity of assumptions.

THEOREM 3A: *If P is a g-prime system with abscissa of convergence α and if $\sum_{j=1}^{\infty} p_j^{-\alpha}$ diverges, then*

$$\liminf_{x \to \infty} \frac{\pi(x)}{I(x)} = 0.$$

Proof: For $\sigma > \alpha$ we have

$$\log \zeta(\sigma) = \sum_{j=1}^{\infty} \sum_{k=1}^{\infty} \frac{1}{k} p_j^{-k\sigma} \geqq \sum_{j=1}^{\infty} p_j^{-\sigma} = \int_1^{\infty} x^{-\sigma} d\pi(x).$$

Integrating the last expression by parts, we obtain

$$\log \zeta(\sigma) \geqq \sigma \int_{p_1}^{\infty} \pi(x) x^{-\sigma-1}\, dx.$$

If the conclusion of the theorem were false, there would exist a positive number c such that $\pi(x)/I(x) \geqq c$ for all $x \geqq p_1$. For $\sigma > \alpha$ we then would have

$$\begin{aligned} \log \zeta(\sigma) &\geqq c\sigma \int_{p_1}^{\infty} I(x) x^{-\sigma-1}\, dx \\ &= c\zeta(\sigma) - c\sigma \int_{1}^{p_1} x^{-\sigma-1}\, dx \\ &> c\zeta(\sigma) - c. \end{aligned}$$

Hence

$$\zeta(\sigma)^{-1} \exp\{c\zeta(\sigma)\} < e^{c} \quad (\sigma > \alpha). \tag{3.1}$$

If $\sigma \downarrow \alpha$, our divergence assumption gives $\zeta(\sigma) \to \infty$. In this case the left-hand side of (3.1) tends to ∞, while the right side is bounded. Thus there exists no positive c such that $\pi(x) \geqq c\, I(x)$ for all $x \geqq p_1$. ▮

We now show that without further hypotheses, one cannot assert that $\lim_{x\to\infty} \pi(x)/I(x) = 0$. (In the next section we shall introduce a certain smoothness assumption with which we can prove that this limit holds.) Since $\pi(x) \leqq I(x)$, we must have $\limsup \pi(x)/I(x) \leqq 1$.

LEMMA 3B: *There is a g-prime system P for which*

$$\limsup_{x\to\infty} \frac{\pi_P(x)}{I_P(x)} = 1.$$

Proof: Let $q_1 = 2$ and $(q_2, q_3, \cdots)$ be an increasing sequence of positive integers to be chosen successively. Let P be the g-prime system obtained by putting the number nq_n in P with multiplicity q_n $(n = 1, 2, \cdots)$, so that $p_j = nq_n$ for

$$q_1 + \cdots + q_{n-1} < j \leqq q_1 + \cdots + q_n.$$

To start, we have $p_1 = p_2 = 2$. It is easy to verify that $\Sigma\, p_j^{-1}$ diverges but $\Sigma p_j^{-1-\epsilon}$ converges for any positive ϵ. Suppose that $q_1, q_2, \cdots, q_{n-1}$ have been chosen. We proceed to choose q_n as follows. Put $m = q_1 + q_2 + \cdots + q_{n-1}$. If $p_1^{\nu_1} \cdots p_m^{\nu_m} \leqq x$, then clearly $\nu_j \leqq \left[\frac{\log x}{\log p_j}\right]$ for $j = 1, 2, \cdots, m$. Thus the generalized integers not exceeding x which can be built up from the first m generalized primes of P are in number at most

$$\left[1 + \frac{\log x}{\log p_1}\right] \cdots \left[1 + \frac{\log x}{\log p_m}\right] \leqq \left(1 + \frac{\log x}{\log 2}\right)^m.$$

For $n > 1$ choose q_n as a positive integer so large that

$$\left(1 + \frac{\log nq_n}{\log 2}\right)^m q_n^{-1/2} < 1.$$

This is possible since $(a + b \log x)^m x^{-1/2}$ approaches zero as $x \to \infty$ with a and b fixed. Now the integers not exceeding nq_n consist of the generalized prime nq_n itself with multiplicity q_n and the integers built up from the first m generalized primes. Hence

$$I(nq_n) \leqq q_n + \left(1 + \frac{\log nq_n}{\log 2}\right)^m < q_n + q_n^{1/2}.$$

On the other hand, $\pi(nq_n) = q_n + q_{n-1} + \cdots + q_1 \geqq q_n$, and so

$$\frac{\pi(nq_n)}{I(nq_n)} > \frac{q_n}{q_n + q_n^{1/2}} = 1 - \frac{1}{q_n^{1/2} + 1}.$$

Thus, $\limsup \pi_P(x)/I_P(x) = 1$. ▮

4. REGULAR GROWTH

We say that a function f has *regular growth* if the following conditions hold: f is a positive nondecreasing function on $[1, \infty)$ and there exists a nonnegative number α such that $\lim_{x\to\infty} f(cx)/f(x) = c^\alpha$ for each positive number c. These conditions are equivalent to requiring that f be positive and nondecreasing and that $f(x)x^{-\alpha}$ be a slowly varying function of x (cf. Section 1). We shall call α the *exponent of growth* associated with f. We shall be interested in the case in which $f = I_P$ for some generalized prime number system P. In this case Lemma 2B and inequality (4.2) (below) imply that α is the abscissa of convergence of P.

LEMMA 4A: *Let I be a function with regular growth and exponent* *of growth* α. *For each* $\epsilon > 0$ *there exist positive numbers* C_ϵ *and* D_ϵ *such that*

$$\frac{1}{C_\epsilon}\left(\frac{x}{y}\right)^{\alpha-\epsilon} \leqq \frac{I(x)}{I(y)} \leqq D_\epsilon\left(\frac{x}{y}\right)^{\alpha+\epsilon} \quad (1 \leqq y \leqq x). \tag{4.1}$$

In particular, if we take $y = 1$, $C'_\epsilon = C_\epsilon/I(1)$, *and* $D'_\epsilon = D_\epsilon I(1)$, *then (for* $x \geqq 1$*)*

$$\frac{1}{C'_\epsilon}x^{\alpha-\epsilon} \leqq I(x) \leqq D'_\epsilon x^{\alpha+\epsilon}. \tag{4.2}$$

REMARKS: *We shall prove only the first inequality in* (4.1), *since the proof of the second one is similar. We have included monotonicity in the definition of a function of regular growth, although the lemma could be proved under weaker hypotheses (cf.* [12]). *Since our functions* I_P *are in fact monotone, we shall content ourselves with the use of this simplifying hypothesis.*

Proof: By hypothesis,

$$\lim_{x\to\infty} I(2x)(2x)^{-\alpha+\epsilon}\{I(x)x^{-\alpha+\epsilon}\}^{-1} = 2^\epsilon.$$

Thus there exists a number $A_\epsilon \geqq 1$ such that

$$I(2z)(2z)^{-\alpha+\epsilon} \geqq I(z)z^{-\alpha+\epsilon} \quad (z \geqq A_\epsilon). \tag{4.3}$$

If $A_\epsilon \leqq y \leqq x$, we put $n = [\log (x/y)/\log 2]$ and apply (4.3) repeatedly:

$$\frac{I(x)x^{-\alpha+\epsilon}}{I(y)y^{-\alpha+\epsilon}} \geqq \frac{I(x)x^{-\alpha+\epsilon}}{I(2y)(2y)^{-\alpha+\epsilon}} \geqq \cdots \geqq \frac{I(x)x^{-\alpha+\epsilon}}{I(2^n y)(2^n y)^{-\alpha+\epsilon}}.$$

Using the monotonicity of I and the fact that $2^n y \leqq x < 2^{n+1}y$, we have

$$\frac{I(x)x^{-\alpha+\epsilon}}{I(y)y^{-\alpha+\epsilon}} \geqq \left(\frac{x}{2^n y}\right)^{-\alpha+\epsilon} \geqq \left(\frac{x}{2^n y}\right)^{-\alpha} \geqq 2^{-\alpha}.$$

If $1 \leqq y \leqq x \leqq A_\epsilon$, then

$$\frac{I(x)x^{-\alpha+\epsilon}}{I(y)y^{-\alpha+\epsilon}} \geqq \left(\frac{x}{y}\right)^{-\alpha+\epsilon} \geqq \left(\frac{x}{y}\right)^{-\alpha} \geqq A_\epsilon^{-\alpha}.$$

If $1 \leqq y \leqq A_\epsilon \leqq x$, then we have

$$\frac{I(x)x^{-\alpha+\epsilon}}{I(y)y^{-\alpha+\epsilon}} = \frac{I(x)x^{-\alpha+\epsilon}}{I(A_\epsilon)A_\epsilon^{-\alpha+\epsilon}}\frac{I(A_\epsilon)A_\epsilon^{-\alpha+\epsilon}}{I(y)y^{-\alpha+\epsilon}} \geqq 2^{-\alpha}A_\epsilon^{-\alpha}$$

by the two previous cases. Our assertion therefore holds with $C_\epsilon = (2A_\epsilon)^\alpha$. ▌

Let P be a g-prime system such that I_P has regular growth. We shall show that the condition of Theorem 3A, namely the divergence of $\sum p_j^{-\alpha}$, implies that $\pi_P(x)/I_P(x) = o(1)$. Our proof uses a modification of the classical "sieve method" of showing $\pi(x)/[x] = o(1)$ for ordinary integers. We shall require the following definition and combinatorial lemma.

Let P be a g-prime system and r a positive integer. Define a function $I_P^{(r)} = I^{(r)}$ on $[1, \infty)$ by taking $I^{(r)}(x)$ to be the number of sequences $(\nu_1, \nu_2, \cdots)$ of integers such that

$$\nu_1 = \cdots = \nu_r = 0; \quad \nu_{r+1} \geqq 0, \quad \nu_{r+2} \geqq 0, \quad \cdots; \quad \text{and} \quad \prod p_j^{\nu_j} \leqq x.$$

Lemma 4B: *Let P be a g-prime system and $I_P^{(r)}$ be defined as above. Then*

$$(4.4) \qquad I_P^{(r)}(x) = \sum_{\delta_1=0}^{1} \cdots \sum_{\delta_r=0}^{1} (-1)^{\delta_1+\cdots+\delta_r} I_P(xp_1^{-\delta_1}\cdots p_r^{-\delta_r}).$$

(This lemma is an instance of Sylvester's principle of cross classification {cf. [22], vol. 2, pp. 119–120 and 326–329}.)

Proof: We observe that

$$p_1^{\mu_1}p_2^{\mu_2} \cdots \leqq xp_1^{-\delta_1}\cdots p_r^{-\delta_r}$$

if and only if

$$p_1^{\mu_1+\delta_1}\cdots p_r^{\mu_r+\delta_r}p_{r+1}^{\mu_{r+1}}p_{r+2}^{\mu_{r+2}} \cdots \leqq x.$$

Thus

$$I_P(xp_1^{-\delta_1}\cdots p_r^{-\delta_r}) = \sum_{\substack{p_1^{\nu_1}p_2^{\nu_2}\cdots \leqq x \\ \nu_1 \geqq \delta_1, \cdots, \nu_r \geqq \delta_r}} 1.$$

Let $J^{(r)}(x)$ denote the right side of (4.4). Then

$$J^{(r)}(x) = \sum_{\delta_1=0}^{1} \cdots \sum_{\delta_r=0}^{1} (-1)^{\delta_1+\cdots+\delta_r} \sum_{\substack{p_1^{\nu_1}p_2^{\nu_2}\cdots \leqq x \\ \nu_1 \geqq \delta_1, \cdots, \nu_r \geqq \delta_r}} 1.$$

Changing the order of summation, we obtain

$$J^{(r)}(x) = \sum_{p_1^{\nu_1}p_2^{\nu_2}\cdots \leqq x} \; \sum_{\delta_1=0}^{\min(\nu_1,1)} \cdots \sum_{\delta_r=0}^{\min(\nu_r,1)} (-1)^{\delta_1+\cdots+\delta_r}.$$

Since

$$\sum_{\delta_j=0}^{\min(\nu_j,1)} (-1)^{\delta_j} = \begin{cases} 0 & \text{if} \quad \nu_j \geqq 1, \\ 1 & \text{if} \quad \nu_j = 0, \end{cases}$$

we have

$$J^{(r)}(x) = \sum_{\substack{p_1^{\nu_1} p_2^{\nu_2} \cdots \leqq x \\ \nu_1 = \cdots = \nu_r = 0}} 1 = I_P^{(r)}(x). \quad ∎$$

THEOREM 4C: *Let P be a g-prime system such that I_P has regular growth and exponent of growth α. Then*

$$\overline{\lim_{x\to\infty}} \frac{\pi_P(x)}{I_P(x)} \leqq \prod_{j=1}^{r} (1 - p_j^{-\alpha})$$

for any positive integer r. In particular, if $\sum p_j^{-\alpha}$ diverges, then

$$\lim_{x\to\infty} \pi_P(x)/I_P(x) = 0.$$

Proof: Let $I^{(r)}$ be the counting function defined above. We have

$$I^{(r)}(x) = \sum_{p_{r+1}^{\nu_{r+1}} p_{r+2}^{\nu_{r+2}} \cdots \leqq x} 1 \geqq \sum_{\substack{p_j \leqq x \\ j > r}} 1 = \pi(x) - r$$

or

$$\pi(x) \leqq r + I^{(r)}(x).$$

Using the last inequality and Lemma 4B, we obtain

$$\overline{\lim_{x\to\infty}} \frac{\pi(x)}{I(x)} \leqq \overline{\lim_{x\to\infty}} \left\{ \frac{r}{I(x)} + \sum_{\delta_1=0}^{1} \cdots \sum_{\delta_r=0}^{1} (-1)^{\delta_1+\cdots+\delta_r} \frac{I(xp_1^{-\delta_1}\cdots p_r^{-\delta_r})}{I(x)} \right\}.$$

Since I is of regular growth, the last expression can be simplified to give

$$\overline{\lim_{x\to\infty}} \frac{\pi(x)}{I(x)} \leqq \sum_{\delta_1=0}^{1} \cdots \sum_{\delta_r=0}^{1} (-1)^{\delta_1+\cdots+\delta_r} (p_1^{-\delta_1}\cdots p_r^{-\delta_r})^{\alpha} = \prod_{j=1}^{r} (1 - p_j^{-\alpha}).$$

The final assertion of the theorem follows from Lemma 2A. ∎

5. A NORMALIZATION: $\alpha = 1$

A generalized prime number system for which the abscissa of convergence α is zero bears very little resemblance to the ordinary prime numbers. A typical example in which $\alpha = 0$ is that in which $p_j = e^j$. In this example the sequence $(n_1, n_2, \cdots)$ consists of the numbers of the form e^k, where k is a positive integer and where e^k occurs with a multiplicity equal to the number of unrestricted partitions of k (cf. [8]). For this reason we sometimes use the term "partition problem" to describe the case $\alpha = 0$. If $0 < \alpha < +\infty$, there is no loss of generality in assuming that α is 1, as the following theorem shows.

THEOREM 5A: *If* $P = (p_1, p_2, \cdots)$ *is a generalized prime number system with abscissa of convergence* $\alpha > 0$ *and if we form a new generalized prime number system* $P' = (p_1', p_2', \cdots)$ *by putting* $p_j' = p_j^\alpha$, *then* P' *has abscissa of convergence* 1 *and we have the following relations between the new system and the old:*

$$n_i' = n_i^\alpha, \quad \pi_{P'}(x^\alpha) = \pi_P(x), \quad I_{P'}(x^\alpha) = I_P(x), \quad \zeta_{P'}(s) = \zeta_P(\alpha s).$$

The proof of each equality is a simple verification. ▌

From now on we shall restrict our attention to generalized prime number systems P *for which the abscissa of convergence is* 1.

6. GENERALIZATION OF RÉNYI'S THEOREM

Let $P = (p_1, p_2, \cdots)$ be a g-prime system. As we indicated earlier, we may enumerate those sequences $(\nu_1, \nu_2, \cdots)$ of nonnegative integers, all but a finite number of which are zero, with the location of a sequence in the enumeration being determined by the size of the product $p_1^{\nu_1} p_2^{\nu_2} \cdots$. Let the ith such sequence be $(\nu_{i1}, \nu_{i2}, \cdots)$ beginning with $(\nu_{01}, \nu_{02}, \cdots) = (0, 0, 0, \cdots)$. Thus

$$n_i = \prod_{j=1}^{\infty} p_j^{\nu_{ij}} \quad (i = 0, 1, 2, \cdots).$$

For $i = 1, 2, \cdots$ let

$$\beta_i = \sum_{j=1}^{\infty} (\nu_{ij} - \operatorname{sgn} \nu_{ij}) = \sum_{j=1}^{\infty} \nu_{ij} - \sum_{j=1}^{\infty} \operatorname{sgn} \nu_{ij}$$

be the "excess" of the ith factorization, that is, the difference between the total number of prime factors counting repetitions and the number of prime factors occurring with nonzero exponent. Here $\operatorname{sgn} x$ is 1 when $x > 0$ and 0 when $x = 0$. We define a function $B_P^{(k)}$ on $[1, \infty)$ by taking $B_P^{(k)}(x)$ to be the number of generalized integers n_i not exceeding x with excess exactly equal to k. That is,

$$B^{(k)}(x) = B_P^{(k)}(x) = \sum_{\substack{n_i \leq x \\ \beta_i = k}} 1.$$

With this notation we have the following theorem.

THEOREM 6A: *If P is a g-prime system with abscissa of convergence* 1 *such that I_P has regular growth, then*

$$\lim_{x \to \infty} \frac{B_P^{(k)}(x)}{I_P(x)} = \rho_k,$$

where $\sum_{k=0}^{\infty} \rho_k z^k$ is the Maclaurin series of the meromorphic function F defined for $z \neq p_1, p_2, \cdots$ by any one of the four expressions

$$\begin{aligned} F(z) &= \prod_{j=1}^{\infty} \left\{ \left(1 - \frac{1}{p_j}\right)\left(1 + \frac{1}{p_j - z}\right) \right\} = \prod_{j=1}^{\infty} \left\{ 1 + \frac{z-1}{p_j(p_j - z)} \right\} \\ &= \frac{1}{\zeta_P(2)} \prod_{j=1}^{\infty} \frac{1 - z/(p_j + 1)}{1 - z/p_j} \\ &= \frac{1}{\zeta_P(2)} \prod_{j=1}^{\infty} \left\{ 1 + \frac{z}{(p_j + 1)(p_j - z)} \right\}. \end{aligned}$$

REMARKS: *Note that $\rho_0 = 1/\zeta_P(2)$, a generalization of the familiar result on the density of the square-free numbers, and that*

$$\rho_0 + \rho_1 + \rho_2 + \cdots = F(1) = 1.$$

In accordance with Theorem 5A, *the above theorem is equivalent to an apparently more general result about g-prime systems with regular growth and positive abscissa of convergence α.*

Before starting to prove the theorem, we give a definition and lemma which we shall presently need. Let P be a g-prime system,

r a fixed positive integer, and $\mu_1, \cdots, \mu_r$ given nonnegative integers. For $x \geqq 1$ define $I_P(x; \mu_1, \cdots, \mu_r)$ to be the number of sequences $(\nu_1, \nu_2, \cdots)$ of nonnegative integers such that $\nu_1 = \mu_1, \cdots, \nu_r = \mu_r$ and $p_1^{\nu_1}p_2^{\nu_2}\cdots \leqq x$. It is clear that $I(x; 0, 0, \cdots, 0) = I^{(r)}(x)$, as defined in Section 4.

LEMMA 6B: *Let P be a g-prime system and let $I_P(x; \mu_1, \cdots, \mu_r)$ be as defined above. Then*

$$I_P(x; \mu_1, \cdots, \mu_r) = \sum_{\delta_1=0}^{1} \cdots \sum_{\delta_r=0}^{1} (-1)^{\delta_1+\cdots+\delta_r} I_P(xp_1^{-\mu_1-\delta_1}\cdots p_r^{-\mu_r-\delta_r}).$$

Proof: The present lemma is a direct generalization of Lemma 4B, and the proof follows at once from Equation (4.4) and the simple relation

$$I_P(x; \mu_1, \cdots, \mu_r) = I_P^{(r)}(xp_1^{-\mu_1}\cdots p_r^{-\mu_r}). \blacksquare$$

Proof of the Theorem: For real u let us write

$$e(u) = e^{2\pi i u} = \cos 2\pi u + i \sin 2\pi u.$$

(Here i is the imaginary unit.) Let k be a given nonnegative integer. By the orthogonality property of the trigonometric functions

$$\frac{B^{(k)}(x)}{I(x)} = \int_0^1 e(-kt)\, \frac{1}{I(x)} \sum_{n_i \leqq x} e(\beta_i t)\, dt. \tag{6.1}$$

The proof now falls naturally into three parts.

First, we show that if r is a sufficiently large positive integer, then the integrand in (6.1) is changed only slightly, uniformly in x and t ($x \geqq 1$, $0 \leqq t \leqq 1$), if we replace

$$\beta_i = \sum_{j=1}^{\infty} (\nu_{ij} - \operatorname{sgn} \nu_{ij})$$

by its truncation

$$\beta_{ir} = \sum_{j=1}^{r} (\nu_{ij} - \operatorname{sgn} \nu_{ij}).$$

Second, we show that if r is fixed, then

$$\lim_{x\to\infty} \frac{1}{I(x)} \sum_{n_i \leqq x} e(\beta_{ir} t) = F_r(e(t)) \tag{6.2}$$

uniformly in t, where

$$F_r(z) = \prod_{j=1}^{r} \left\{\left(1 - \frac{1}{p_j}\right)\left(1 + \frac{1}{p_j - z}\right)\right\} = \prod_{j=1}^{r} \left\{1 + \frac{z-1}{p_j(p_j - z)}\right\}.$$

Third, we note that as r tends to infinity, $F_r \to F$ uniformly in the closed unit disc, and so, if r is sufficiently large, $|F(e(t)) - F_r(e(t))|$ is uniformly small for $0 \leqq t \leqq 1$. The three parts of the proof are combined as follows:

By taking r so large that the discrepancies in the first and third parts of the proof are as small as desired, we deduce from (6.2) that

$$\lim_{x\to\infty} \frac{1}{I(x)} \sum_{n_i \leqq x} e(\beta_i t) = F(e(t))$$

uniformly in $t(0 \leqq t \leqq 1)$. This last equation gives

$$\int_0^1 \lim_{x\to\infty} \frac{1}{I(x)} \sum_{n_i \leqq x} e(\beta_i t)e(-kt)\, dt = \int_0^1 F(e(t))e(-kt)\, dt. \tag{6.3}$$

By the Cauchy integral formula, the right side of (6.3) is ρ_k, the kth coefficient of the Maclaurin series of F. The limit on the left side of (6.3) is uniform in t, and thus we may take the limit outside the integral. Taking limits in (6.1), we obtain

$$\begin{aligned} \lim_{x\to\infty} \frac{B^{(k)}(x)}{I(x)} &= \lim_{x\to\infty} \int_0^1 e(-kt)\, \frac{1}{I(x)} \sum_{n_i \leqq x} e(\beta_i t)\, dt \\ &= \int_0^1 F(e(t))e(-kt)\, dt = \rho_k. \end{aligned}$$

We now give the details.

1. If ξ and η are real, then

$$\begin{aligned} |e(\xi) - e(\eta)| &= \left| e\left(\frac{\xi - \eta}{2}\right) - e\left(\frac{\eta - \xi}{2}\right)\right| \\ &= |2 \sin \pi(\xi - \eta)| \leqq 2\pi|\xi - \eta|. \end{aligned}$$

Thus if $0 \leqq t \leqq 1$, we have

$$\begin{aligned} \left|\frac{1}{I(x)} \sum_{n_i \leqq x} e(\beta_i t) - \frac{1}{I(x)} \sum_{n_i \leqq x} e(\beta_{ir} t)\right| &\leqq \frac{2\pi}{I(x)} \sum_{n_i \leqq x} (\beta_i - \beta_{ir}) \\ &= \frac{2\pi}{I(x)} \sum_{j=r+1}^{\infty} \sum_{n_i \leqq x} (\nu_{ij} - \operatorname{sgn} \nu_{ij}). \end{aligned} \tag{6.4}$$

For fixed j, we group equal values of the inner summand in the last expression:

$$\sum_{n_i \leqq x} (\nu_{ij} - \operatorname{sgn} \nu_{ij}) = \sum_{m=2}^{\infty} (m-1) \sum_{\substack{n_i \leqq x \\ \nu_{ij} = m}} 1. \tag{6.5}$$

For fixed j and m, with $m \geqq 2$, we have, by an argument analogous to that used in proving Lemma 6B,

$$\sum_{\substack{n_i \leqq x \\ \nu_{ij} = m}} 1 = \sum_{\substack{n_i \leqq x \\ \nu_{ij} \geqq m}} 1 - \sum_{\substack{n_i \leqq x \\ \nu_{ij} > m}} 1 = I(xp_j^{-m}) - I(xp_j^{-m-1}). \tag{6.6}$$

We combine (6.5) and (6.6) and sum by parts:

$$\begin{aligned} \sum_{n_i \leqq x} (\nu_{ij} - \operatorname{sgn} \nu_{ij}) &= \sum_{m=2}^{\infty} (m-1) \left\{ I\left(\frac{x}{p_j^m}\right) - I\left(\frac{x}{p_j^{m+1}}\right) \right\} \\ &= \sum_{m=2}^{\infty} I\left(\frac{x}{p_j^m}\right). \end{aligned} \tag{6.7}$$

This operation is valid since $I(xp_j^{-m}) = 0$ for m sufficiently large, and thus there are no "end terms" from the summation by parts.

The terms $I(xp_j^{-m})$ may be estimated by using Lemma 4A with $\epsilon = \frac{1}{3}$ and $C = C_{1/3}$. For $p_j^{-m}x \geqq 1$, we have

$$I(p_j^{-m}x) \leqq C(p_j^{-m})^{2/3} I(x). \tag{6.8}$$

For $p_j^{-m}x < 1$, the left side of (6.8) is zero and thus the inequality holds in this case as well. By using the relations (6.4), (6.7), and (6.8), we find

$$\begin{aligned} \left| \frac{1}{I(x)} \sum_{n_i \leqq x} e(\beta_i t) - \frac{1}{I(x)} \sum_{n_i \leqq x} e(\beta_{ir} t) \right| &\leqq 2\pi \sum_{j=r+1}^{\infty} \sum_{m=2}^{\infty} C p_j^{-2m/3} \\ &\leqq \frac{2\pi C}{1 - p_1^{-2/3}} \sum_{j=r+1}^{\infty} p_j^{-4/3} \end{aligned} \tag{6.9}$$

uniformly in x and t for $x \geqq 1$ and $0 \leqq t \leqq 1$. By choosing r suitably large we can make the final quantity in (6.9) as small as desired.

2. We write

$$\begin{aligned} \frac{1}{I(x)} \sum_{n_i \leqq x} e(\beta_{ir} t) &= \frac{1}{I(x)} \sum_{n_i \leqq x} e\left(\sum_{j=1}^{r} \{\nu_{ij} - \operatorname{sgn} \nu_{ij}\} t \right) \\ &= \frac{1}{I(x)} \sum_{\mu_1=0}^{\infty} \cdots \sum_{\mu_r=0}^{\infty} e\left(\sum_{j=1}^{r} \{\mu_j - \operatorname{sgn} \mu_j\} t \right) \cdot \sum_{\substack{n_i \leqq x \\ \nu_{i1} = \mu_1, \cdots, \nu_{ir} = \mu_r}} 1. \end{aligned}$$

Recall that

$$\sum_{\substack{n_i \leqq x \\ \nu_{i1}=\mu_1, \cdots, \nu_{ir}=\mu_r}} 1 = I(x; \mu_1, \cdots, \mu_r)$$

by definition. We evaluate $I(x; \mu_1, \cdots, \mu_r)$ by Lemma 6B and now write

$$\frac{1}{I(x)} \sum_{n_i \leqq x} e(\beta_{ir}t) = \sum_{\mu_1=0}^{\infty} \cdots \sum_{\mu_r=0}^{\infty} e\left(\sum_{j=1}^{r} \{\mu_j - \operatorname{sgn} \mu_j\} t\right)$$
$$\cdot \sum_{\delta_1=0}^{1} \cdots \sum_{\delta_r=0}^{1} (-1)^{\delta_1 + \cdots \delta_r} \frac{I(x p_1^{-\mu_1-\delta_1} \cdots p_r^{-\mu_r-\delta_r})}{I(x)}.$$

Lemma 4A, with $\epsilon = \frac{1}{3}$, shows the last sum to be majorized by

$$\sum_{\mu_1=0}^{\infty} \cdots \sum_{\mu_r=0}^{\infty} \sum_{\delta_1=0}^{1} \cdots \sum_{\delta_r=0}^{1} C_{1/3} (p_1^{\mu_1+\delta_1} \cdots p_r^{\mu_r+\delta_r})^{-2/3}$$

independently of x and t; thus that sum converges uniformly in x and t. Since $\lim_{x \to \infty} I(cx)/I(x) = c$ for any fixed positive c, we have

$$\lim_{x \to \infty} \frac{1}{I(x)} \sum_{n_i \leqq x} e(\beta_{ir}t) = \sum_{\mu_1=0}^{\infty} \cdots \sum_{\mu_r=0}^{\infty} e\left(\sum_{j=1}^{r} \{\mu_j - \operatorname{sgn} \mu_j\} t\right)$$
$$\cdot p_1^{-\mu_1} \cdots p_r^{-\mu_r} (1 - p_1^{-1}) \cdots (1 - p_r^{-1})$$
$$= \prod_{j=1}^{r} \left\{(1 - p_j^{-1}) \left(1 + \frac{1}{p_j - e(t)}\right)\right\} = F_r(e(t))$$

uniformly in t, provided r is fixed. This establishes (6.2).

3. If $|\alpha_j| \leqq a_j$ for $j = 1, 2, \cdots$, we have

$$\left|\prod_{j=1}^{n} (1 + \alpha_j) - \prod_{j=1}^{r} (1 + \alpha_j)\right| \leqq \prod_{j=1}^{n} (1 + a_j) - \prod_{j=1}^{r} (1 + a_j)$$

for $n \geqq r$. If in addition $\prod_{j=1}^{\infty} (1 + a_j)$ converges, so does $\prod_{j=1}^{\infty} (1 + \alpha_j)$ and the inequality

$$\left|\prod_{j=1}^{\infty} (1 + \alpha_j) - \prod_{j=1}^{r} (1 + \alpha_j)\right| \leqq \prod_{j=1}^{\infty} (1 + a_j) - \prod_{j=1}^{r} (1 + a_j) \tag{6.10}$$

is valid.

Now we have the simple inequalities

$$\left|\frac{z-1}{p_j(p_j-z)}\right| \leqq \frac{2}{p_j(p_j-1)} \leqq \frac{2p_1}{p_1-1}\frac{1}{p_j^2} \qquad (|z| \leqq 1)$$

and

$$\prod_{j=1}^{\infty}\left(1+\frac{2p_1}{(p_1-1)p_j^2}\right) \leqq \prod_{j=1}^{\infty} \exp \frac{2p_1}{(p_1-1)p_j^2}$$
$$= \exp\left\{\frac{2p_1}{p_1-1}\sum_{j=1}^{\infty} p_j^{-2}\right\} < \infty.$$

Using (6.10) with

$$\alpha_j = \frac{z-1}{p_j(p_j-z)} \quad \text{and} \quad a_j = \frac{2p_1}{p_1-1}p_j^{-2},$$

we obtain

(6.11) $|F(z) - F_r(z)|$

$$\leqq \prod_{j=1}^{\infty}\left\{1+\frac{2p_1}{p_1-1}p_j^{-2}\right\} - \prod_{j=1}^{r}\left\{1+\frac{2p_1}{p_1-1}p_j^{-2}\right\}$$

uniformly for $|z| \leqq 1$. By choosing r suitably large, we can make the right-hand side of (6.11) as small as desired.

Combining the three parts of the proof as we have indicated earlier, we obtain the assertion of the theorem. ∎

7. THE PRIME NUMBER THEOREM FOR GENERALIZED NUMBERS

We now turn to Beurling's theorem on the distribution of g-primes. For this we must assume, as we indicated in Section 1, that P is a g-prime system whose integer counting function I_P satisfies more stringent conditions than regular growth.

THEOREM 7A (Beurling): *Let P be a generalized prime number system with the property that there exist numbers $A > 0$ and γ such that*

$$I_P(x) = Ax + O(x \log^{-\gamma} x). \tag{7.1}$$

If $\gamma > \frac{3}{2}$, then $\pi(x) = \dfrac{x}{\log x} + o\left(\dfrac{x}{\log x}\right)$.

The next five sections will be devoted to the proof of this theorem. The main steps in the proof consist of (1) showing that $\zeta_P(s) \neq 0$ for $\text{Re } s \geqq 1$, (2) obtaining information on the behavior of ζ' near the line $\text{Re } s = 1$, and (3) extracting an asymptotic estimate of the ψ function (cf. Lemma 2D) from $\zeta'(s)/\zeta(s)$ by a tauberian theorem.

Our arguments are frequently more delicate than those needed to prove the prime number theorem for ordinary numbers. The need for this delicacy stems from the fact that for $\gamma < 2$, ζ' need not have a limit or even be bounded as its argument tends to a point $1 + i\tau$ lying on the line of convergence. For $\gamma > 2$, these difficulties largely vanish, and we could adapt most classical methods which do not depend on the behavior of ζ for $\text{Re } s < 1$.

REMARK: *If* (7.1) *holds for* I_P *with some number* γ, *then* (7.1) *holds a fortiori for a number* $\gamma' < \gamma$. *We shall use this observation later to avoid special cases by assuming that* $\gamma < 2$.

Of course, $I_P(x)$ is bounded near $x = 1$. Therefore, by (7.1) we can find a number $B = B_P \geqq 1$ such that

$$|I_P(x) - Ax| \leqq Bx\,(\log ex)^{-\gamma} \quad (x \geqq 1). \tag{7.2}$$

We have introduced the e to obtain an estimation function which is also bounded near 1.

We begin our work by showing that we can extend ζ to a continuous function on $\{s\colon \text{Re } s \geqq 1,\ s \neq 1\}$.

LEMMA 7B: *Let* P *satisfy* (7.1) *with* $\gamma > 1$ *and let* ζ *be the associated zeta function. There is a function* g *defined by*

$$g(s) = \int_1^\infty x^{-s-1}\{I(x) - Ax\}\,dx$$

which is continuous on $\text{Re } s \geq 1$ *and on* $\text{Re } s > 1$ *satisfies*

$$\zeta(s) = \frac{A}{s-1} + A + sg(s). \tag{7.3}$$

As a consequence of this lemma, ζ has a limit as its argument approaches $1 + it \neq 1$ through values s with $\text{Re } s > 1$, and *we shall define* $\zeta(1 + it)$, $t \neq 0$, *by Equation* (7.3).

Proof of Lemma 7B. We modify Equation (2.3) to obtain

$$\zeta(s) = s \int_1^\infty x^{-s} \left\{ \frac{I(x) - Ax}{x} \right\} dx + A + \frac{A}{s-1}. \tag{7.4}$$

By (7.2), we have the bound $B(\log ex)^{-\gamma}$ for the expression in braces. Since $\gamma > 1$, the integral in (7.4) converges uniformly for $\operatorname{Re} s \geq 1$, and thus the function g is continuous on this set. ∎

8. ZETA HAS NO ZEROS ON $\operatorname{Re} s = 1$

We know from Lemma 2C that ζ has no zeros on $\{s\colon \operatorname{Re} s > 1\}$. If (7.1) holds with $\gamma > \frac{3}{2}$, the zero-free region may be extended to include the line $\operatorname{Re} s = 1$. We shall show this by proving on the one hand that the functional values of ζ cannot tend to zero too swiftly as the argument of ζ tends to $1 + it$, and on the other hand that ζ satisfies a certain Lipschitz condition. These two facts will imply that $\zeta(1 + it) \neq 0$.

We begin by introducing the Fejér kernel K_N. For $N = 1, 2, \cdots$, define the function K_N on the reals by the formula

$$NK_N(x) = N + \sum_{j=1}^{N} 2(N - j) \cos jx.$$

LEMMA 8A: *For each N, K_N assumes only nonnegative values.*

Proof: We have

$$\begin{aligned} 0 &\leqq \left| \sum_{j=1}^{N} e^{ijx} \right|^2 = \left(\sum_{j=1}^{N} e^{ijx} \right) \left(\sum_{k=1}^{N} e^{-ikx} \right) = \sum_{j=1}^{N} \sum_{k=1}^{N} e^{i(j-k)x} \\ &= Ne^0 + (N-1)(e^{ix} + e^{-ix}) + \cdots \\ &\qquad\qquad + (N - (N-1))(e^{i(N-1)x} + e^{-i(N-1)x}) \\ &= NK_N(x). \end{aligned}$$ ∎

We use the Fejér kernel to obtain an inequality for the zeta function.

LEMMA 8B: *For $N = 1, 2, \cdots$ and any real number t, the following inequality is valid:*

$$|\zeta(\sigma)|^N \prod_{l=1}^{N} |\zeta(\sigma + ilt)|^{2N-2l} \geqq 1 \quad (\sigma > 1).$$

Proof: We must show that

$$N \log \zeta(\sigma) + \sum_{l=1}^{N} 2(N - l) \log |\zeta(\sigma + ilt)| \geqq 0. \tag{8.1}$$

Recall from (2.5) that for $\sigma > 1$ we have

$$\zeta(\sigma + it) = \exp \sum_{j=1}^{\infty} \sum_{k=1}^{\infty} k^{-1} p_j^{-k(\sigma+it)}$$

$$= \exp \sum_{j=1}^{\infty} \sum_{k=1}^{\infty} k^{-1} p_j^{-k\sigma} \{\cos(tk \log p_j) - i \sin(tk \log p_j)\}.$$

Thus

$$\log |\zeta(\sigma + it)| = \sum_{j=1}^{\infty} \sum_{k=1}^{\infty} k^{-1} p_j^{-k\sigma} \cos(tk \log p_j).$$

The left side of (8.1) may be written as

$$\sum_{j=1}^{\infty} \sum_{k=1}^{\infty} k^{-1} p_j^{-k\sigma} \left\{N + \sum_{l=1}^{N} 2(N - l) \cos (tlk \log p_j)\right\}.$$

The expression in braces is equal to $NK_N(tk \log p_j)$, which is nonnegative, as we have just seen. Thus (8.1) is valid and the lemma is established. ▌

Now we are in a position to show that $\zeta(s)$ cannot tend to zero too quickly as $s \to 1 + it$.

LEMMA 8C: *Let P be a g-prime system for which* (7.1) *holds with* $\gamma > 1$. *Let t be a nonzero real number and N any positive integer. There exists a positive number $C = C_P(N, t)$ such that for* $1 < \sigma \leqq 2$,

$$|\zeta(\sigma + it)| \geqq C(\sigma - 1)^{1/2+1/(2N)}.$$

Proof: We require some simple estimates for ζ stemming from the representation of Lemma 7B and inequality (7.2). The function g is uniformly bounded for $\operatorname{Re} s \geqq 1$ by

$$\int_1^\infty x^{-1} B(\log ex)^{-\gamma}\, dx = \frac{B}{(\gamma - 1)}.$$

For $1 < \sigma \leqq 2$ and $A' = 2A + 2B/(\gamma - 1)$, we have

$$|\zeta(\sigma)| \leqq A'(\sigma - 1)^{-1}.$$

For $1 < \sigma \leqq 2$ and $2 \leqq j \leqq N + 1$, we have

$$|\zeta(\sigma + ijt)| \leqq \frac{A}{2|t|} + A + \{2 + (N + 1)|t|\} \frac{B}{\gamma - 1} = A''.$$

Now we use Lemma 8B with $N + 1$ in place of N and the last two inequalities. We obtain

$$\begin{aligned} 1 &\leqq |\zeta(\sigma)|^{N+1}|\zeta(\sigma + it)|^{2N}|\zeta(\sigma + 2it)|^{2N-2}\cdots|\zeta(\sigma + Nit)|^2 \\ &\leqq \left(\frac{A'}{\sigma - 1}\right)^{N+1} |\zeta(\sigma + it)|^{2N}(A'')^{2+4+\cdots+2N-2}. \end{aligned}$$

Taking $C = (A')^{-(N+1)/(2N)}(A'')^{-(N-1)/2}$, we have

$$|\zeta(\sigma + it)| \geqq C(\sigma - 1)^{(1/2)+1/(2N)}. \quad \blacksquare$$

Now we show that ζ satisfies a Lipschitz condition. For the sake of convenience, we assume, as we may, that $\gamma < 2$. The other case would require some alterations of the lemma.

LEMMA 8D: *Let P be a g-prime system for which* (7.1) *holds with* $1 < \gamma < 2$. *Let δ and Δ be any numbers satisfying*

$$0 < \delta < \Delta - 1 < \infty.$$

Let

$$\Omega = \Omega(\delta, \Delta) = \{s\colon \operatorname{Re} s \geqq 1, |s - 1| \geqq \delta, |s| \leqq \Delta\}.$$

There is a number $D = D_P(\delta, \Delta)$ such that if $s_1 \in \Omega$ and $s_2 \in \Omega$, then

$$|\zeta(s_1) - \zeta(s_2)| \leqq D|s_1 - s_2|^{\gamma-1}.$$

Proof: We use the representation of ζ from Lemma 7B

$$\begin{aligned} (8.2) \quad \zeta(s_1) - \zeta(s_2) &= \frac{A}{s_1 - 1} - \frac{A}{s_2 - 1} \\ &\quad + \int_1^\infty (s_1x^{-s_1} - s_2x^{-s_2})x^{-1}\{I(x) - Ax\}\, dx \end{aligned}$$

and estimate $s_1x^{-s_1} - s_2x^{-s_2}$ differently depending on whether $|s_1 - s_2| \log x$ is smaller or larger than 1. In the first case, we shall apply the estimate

$$\begin{aligned} |s_1x^{-s_1} - s_2x^{-s_2}| &= \left|\int_{s_1}^{s_2} \frac{d}{dz}(zx^{-z})\, dz\right| \\ &\leqq |s_1 - s_2|(1 + \Delta \log x)x^{-1} \\ &\leqq |s_1 - s_2|x^{-1}\, \Delta \log ex, \end{aligned}$$

while in the second case we shall apply the estimate

$$|s_1 x^{-s_1} - s_2 x^{-s_2}| \leqq 2\Delta x^{-1}.$$

Take $X = \exp(|s_1 - s_2|^{-1})$ and write the integral in (8.2) as $\int_1^X + \int_X^\infty$. Using (7.2) and the estimates for $s_1 x^{-s_1} - s_2 x^{-s_2}$, we find that

$$\left|\int_1^X\right| \leqq |s_1 - s_2| \int_1^X B\Delta(\log ex)^{1-\gamma} x^{-1}\, dx$$

$$\leqq B\Delta|s_1 - s_2| \int_1^X (\log x)^{1-\gamma} \frac{dx}{x} = \frac{B\Delta}{2-\gamma} |s_1 - s_2|^{\gamma-1}$$

and

$$\left|\int_X^\infty\right| \leqq 2B\Delta \int_X^\infty (\log ex)^{-\gamma} \frac{dx}{x} = \frac{2B\Delta(\log eX)^{1-\gamma}}{\gamma - 1}$$

$$\leqq \frac{2B\Delta}{\gamma - 1} |s_1 - s_2|^{\gamma-1}.$$

It remains to estimate $A(s_1 - 1)^{-1} - A(s_2 - 1)^{-1}$. If $|s_1 - s_2| \leqq 1$, we have

$$\left|\frac{A}{s_1 - 1} - \frac{A}{s_2 - 1}\right| = \frac{A|s_1 - s_2|}{|s_1 - 1||s_2 - 1|} \leqq \frac{A|s_1 - s_2|^{\gamma-1}}{\delta^2}.$$

If $|s_1 - s_2| > 1$, we have

$$\frac{A|s_1 - s_2|}{|s_1 - 1||s_2 - 1|} \leqq \frac{A(|s_1 - 1| + |s_2 - 1|)}{|s_1 - 1||s_2 - 1|} \leqq \frac{2A}{\delta} \leqq \frac{2A}{\delta} |s_1 - s_2|^{\gamma-1}.$$

Taking

$$D = B\Delta\left(\frac{1}{2-\gamma} + \frac{2}{\gamma - 1}\right) + \frac{A}{\delta}\left(2 + \frac{1}{\delta}\right),$$

we have the desired inequality. ∎

The last two lemmas will now be applied to show that ζ has no zeros if $\gamma > \frac{3}{2}$.

THEOREM 8E: *Let P be a g-prime system satisfying*

$$I_P(x) = Ax + O(x \log^{-\gamma} x) \tag{7.1'}$$

and let ζ_P be the associated zeta function. If $\gamma > \frac{3}{2}$, ζ_P has no zeros in the closed half plane Re $s \geqq 1$.

REMARK: *The condition $\gamma > \frac{3}{2}$ in the above theorem is "sharp" in the sense that the theorem may be false for $\gamma \leqq \frac{3}{2}$. This is in fact the case in the examples mentioned after Equation* (1.6).

Proof: We have shown in Lemma 2C that ζ_P has no zeros in $\operatorname{Re} s > 1$. We shall assume that $\zeta(1 + it_0) = 0$ for some $t_0 \neq 0$ and show that this leads to a contradiction.

Taking $\zeta(1 + it_0) = 0$, we have from Lemma 8D that

$$|\zeta(\sigma + it_0)| \leqq D(\sigma - 1)^{\gamma-1} \quad (1 < \sigma < 2).$$

Let N be an integer chosen so large that $(2N)^{-1} < \gamma - \frac{3}{2}$. We apply Lemma 8C with this value of N to obtain

$$|\zeta(\sigma + it_0)| \geqq C(\sigma - 1)^{(1/2)+(1/2N)} \quad (1 < \sigma < 2).$$

These inequalities together imply that

$$C(\sigma - 1)^{(1/2)+(1/2N)} \leqq D(\sigma - 1)^{\gamma-1} \quad (1 < \sigma < 2)$$

or

$$\frac{C}{D} \leqq (\sigma - 1)^{\gamma-(3/2)-(1/2N)} \quad (1 < \sigma < 2).$$

This last inequality is certainly absurd, since the right side tends to zero as $\sigma \downarrow 1$. Thus ζ can have no zeros on $\operatorname{Re} s \geqq 1$. ∎

9. BEHAVIOR OF ζ'/ζ NEAR $\operatorname{Re} s = 1$

We shall soon need information about ζ'/ζ near the line $\operatorname{Re} s = 1$. We have just shown that the factor $1/\zeta$ is relatively harmless. We form ζ' by differentiating both sides of Equation (7.3).

$$\zeta'(s) = \frac{-A}{(s-1)^2} + \int_1^\infty x^{-s-1}(1 - s \log x)\{I(x) - Ax\}\, dx \quad (\operatorname{Re} s > 1). \tag{9.1}$$

This equation is valid since the integral on the right side converges uniformly on each compact subset of $\operatorname{Re} s > 1$. As we observed before, ζ' need not have a pointwise limit for values of s on the line $\operatorname{Re} s = 1$. However, we shall show ζ' to be well behaved in the mean.

In order to "neutralize" the singularity of ζ'/ζ at $s = 1$, we define a new function on $\operatorname{Re} s > 1$ by setting

$$h(s) = -\frac{1}{s}\frac{\zeta'(s)}{\zeta(s)} - \frac{1}{s-1}.$$

LEMMA 9A: *Assume* (7.1) *holds with* $\gamma > \frac{3}{2}$. *If* λ *is any fixed positive number, then*

$$\lim_{\sigma,\sigma' \downarrow 1} \int_{-\lambda}^{\lambda} |h(\sigma + it) - h(\sigma' + it)|^2 \, dt = 0.$$

REMARK: *Assume for the moment the truth of this lemma and the completeness of* $L^2(-\lambda, \lambda)$, *the space of Lebesgue square integrable functions defined on* $(-\lambda, \lambda)$. *Then there exists a function* $h_1 \in L^2(-\lambda, \lambda)$ *such that*

$$\lim_{\sigma \downarrow 1} \int_{-\lambda}^{\lambda} |h(\sigma + it) - h_1(t)|^2 \, dt = 0.$$

Thus, for each $\lambda > 0$, *we could define*

$$-\frac{\zeta'(1 + it)}{\zeta(1 + it)} - \frac{1 + it}{it} \quad \text{on} \quad -\lambda \leqq t \leqq \lambda$$

as a limit of square integrable functions. However, we do not wish to appeal to the completeness of L^2, *and we shall carry out our subsequent arguments without using this remark.*

Proof: We have

$$h(s) = -\frac{1}{s} - \frac{1}{s^2} - \frac{g(s)}{(s-1)\zeta(s)} - \frac{g'(s)}{\zeta(s)}, \tag{9.2}$$

where g was defined in Lemma 7B. We may verify this equality by noting that

$$(s-1)g(s) = \frac{(s-1)\zeta(s) - As}{s} = \left(1 - \frac{1}{s}\right)\zeta(s) - A$$

and writing

$$-\frac{\{g(s) + (s-1)g'(s)\}}{(s-1)\zeta(s)} = -\frac{\frac{d}{ds}\{(s-1)g(s)\}}{(s-1)\zeta(s)}$$

$$= -\frac{(1 - s^{-1})\zeta'(s) + s^{-2}\zeta(s)}{(s-1)\zeta(s)}$$

$$= -\frac{1}{s}\frac{\zeta'(s)}{\zeta(s)} - \frac{1}{s-1} + \frac{1}{s-1}(1 - s^{-2})$$

$$= h(s) + \frac{1}{s} + \frac{1}{s^2}.$$

Writing $s = \sigma + it$ $(\sigma > 1)$, we set

$$-\frac{1}{s} - \frac{1}{s^2} - \frac{g(s)}{(s-1)\zeta(s)} = J_\sigma(t),$$

$$-\frac{1}{\zeta(s)} = K_\sigma(t),$$

$$g'(s) = L_\sigma(t).$$

To prove the assertion of the lemma, we shall show that J and K can be extended as continuous functions on the half plane $\sigma \geqq 1$ and that L satisfies certain integral conditions.

We have already shown in Lemma 7B that g is continuous on $\operatorname{Re} s \geqq 1$. We have

$$(s-1)\zeta(s) = As + s(s-1)g(s) \quad (\operatorname{Re} s > 1)$$

and thus $s \mapsto (s-1)\zeta(s)$ is a continuous function on $\operatorname{Re} s > 1$ and may be extended to a continuous function on the closed set $\operatorname{Re} s \geqq 1$. We have also shown that $\zeta(s) \neq 0$ on $\operatorname{Re} s \geqq 1$, and thus $(s-1)\zeta(s) \neq 0$ for $\operatorname{Re} s \geqq 1$, with the possible exception of $s = 1$. But $\lim_{s\to 1} (s-1)\zeta(s) = A \neq 0$. Thus the function $s \mapsto \{(s-1)\zeta(s)\}^{-1}$ is continuous on $\operatorname{Re} s \geqq 1$, and hence $\sigma + it \mapsto J_\sigma(t)$ is continuous on $\sigma \geqq 1$. For K, we note that $1/\zeta(s) = (s-1)/\{(s-1)\zeta(s)\}$, $s \neq 1$, and each side has limit zero at $s = 1$. If we set $K_1(0) = 0$, then $\sigma + it \mapsto K_\sigma(t)$ is continuous on $\operatorname{Re} s \geqq 1$. Thus J and K may be extended as uniformly continuous functions on the compact set

$$1 \leqq \sigma \leqq 2, \qquad -\lambda \leqq t \leqq \lambda.$$

In terms of the new symbols, (9.2) may be rewritten as

$$h(s) = J_\sigma(t) + K_\sigma(t)\, L_\sigma(t),$$

and we have for $\sigma, \sigma' > 1$

$$\int_{-\lambda}^{\lambda} |h(\sigma + it) - h(\sigma' + it)|^2 \, dt =$$

$$\int_{-\lambda}^{\lambda} |(J_\sigma - J_{\sigma'}) + (K_\sigma - K_{\sigma'})L_\sigma + K_{\sigma'}(L_\sigma - L_{\sigma'})|^2.$$

By Cauchy's inequality we have

$$|X + Y + Z|^2 \leqq 3(|X|^2 + |Y|^2 + |Z|^2), \tag{9.3}$$

and so

$$\int_{-\lambda}^{\lambda} |h(\sigma + it) - h(\sigma' + it)|^2 \, dt \leqq \tag{9.4}$$

$$3\int_{-\lambda}^{\lambda} |J_\sigma - J_{\sigma'}|^2 + 3\int_{-\lambda}^{\lambda} |(K_\sigma - K_{\sigma'})L_\sigma|^2 + 3\int_{-\lambda}^{\lambda} |K_{\sigma'}(L_\sigma - L_{\sigma'})|^2.$$

The uniform continuity of J implies that the first term on the right side of (9.4) tends to zero as $\sigma, \sigma' \downarrow 1$. The next term on the right side of (9.4) does not exceed

$$3 \max_{-\lambda \leqq t \leqq \lambda} |K_\sigma(t) - K_{\sigma'}(t)|^2 \int_{-\lambda}^{\lambda} |L_\sigma|^2.$$

This expression will tend to zero as $\sigma, \sigma' \downarrow 1$ provided $\int_{-\lambda}^{\lambda} |L_\sigma|^2$ is uniformly bounded. The third term on the right side of (9.4) is bounded by

$$3 \max_{\substack{-\lambda \leqq t \leqq \lambda \\ 1 \leqq \sigma' \leqq 2}} |K_{\sigma'}(t)|^2 \int_{-\lambda}^{\lambda} |L_\sigma - L_{\sigma'}|^2.$$

To complete the proof of Lemma 9A we now estimate

$$\int_{-\lambda}^{\lambda} |L_\sigma|^2 \quad \text{and} \quad \int_{-\lambda}^{\lambda} |L_\sigma - L_{\sigma'}|^2$$

by Plancherel's theorem or the Bessel-type inequality for Fourier transforms which we have proved in Appendix 4.

We have

$$L_\sigma(t) = g'(\sigma + it) = -\int_1^\infty x^{-s-1} \log x \{I(x) - Ax\} \, dx.$$

If we take $x = e^u$ and for each real number σ define a function φ_σ on $[0, \infty)$ by

$$\varphi_\sigma(u) = -e^{-\sigma u} u \{I(e^u) - Ae^u\},$$

we obtain the Fourier integral

$$L_\sigma(t) = \int_0^\infty e^{-itu}\varphi_\sigma(u)\,du \quad (\sigma > 1). \tag{9.5}$$

For each number $\sigma > 1$, φ_σ satisfies the hypotheses of Appendix 4, and thus

$$\int_{-\lambda}^{\lambda} |L_\sigma(t)|^2\,dt \leqq 2\pi\int_0^\infty |\varphi_\sigma(u)|^2\,du.$$

Also, for each $\sigma > 1$ we have by (7.2)

$$|\varphi_\sigma(u)| \leqq |\varphi_1(u)| \leqq B(1 + u)^{1-\gamma}.$$

Together, these inequalities imply that

$$\int_{-\lambda}^{\lambda} |L_\sigma(t)|^2\,dt \leqq 2\pi \int_0^\infty |\varphi_1(u)|^2\,du \leqq 2\pi B^2 \int_0^\infty (1 + u)^{2-2\gamma}\,du < \infty$$

uniformly for all $\sigma > 1$. Thus the second term on the right side of (9.4) tends to zero as $\sigma, \sigma' \downarrow 1$.

Next, we write (9.5) for $L_{\sigma'}(t)$ and subtract, obtaining

$$L_\sigma(t) - L_{\sigma'}(t) = \int_0^\infty e^{-itu}\{\varphi_\sigma(u) - \varphi_{\sigma'}(u)\}\,du.$$

By Appendix 4, we have for $\sigma, \sigma' > 1$

$$\int_{-\lambda}^{\lambda} |L_\sigma(t) - L_{\sigma'}(t)|^2\,dt \leqq 2\pi \int_0^\infty |\varphi_\sigma(u) - \varphi_{\sigma'}(u)|^2\,du. \tag{9.6}$$

Now $|\varphi_\sigma - \varphi_{\sigma'}|^2$ is dominated by the integrable function $|\varphi_1|^2$ and tends to zero uniformly on any fixed bounded interval as $\sigma, \sigma' \downarrow 1$. Thus, given $\epsilon > 0$, there exists a number $X \geqq 0$ such that for all $\sigma, \sigma' > 1$ we have

$$2\pi\int_X^\infty |\varphi_\sigma(u) - \varphi_{\sigma'}(u)|^2\,du < \frac{\epsilon}{2}.$$

For σ and σ' sufficiently close to 1 and for any $u \in [0, X]$ we have $2\pi|\varphi_\sigma(u) - \varphi_{\sigma'}(u)|^2 < \epsilon/2X$. Hence the right side of (9.6) is less than ϵ for σ, σ' close enough to 1. Thus the third term on the right side of (9.4) also has limit zero, and the proof of the lemma is complete.

10. A TAUBERIAN THEOREM

Our plan has been to extract information about ψ from the formula

$$-\frac{1}{s}\frac{\zeta'(s)}{\zeta(s)} = \int_1^\infty x^{-s-1}\psi(x)\,dx,$$

which involves an integral average of the function ψ. A process that uses information about an average of a function to deduce knowledge about the function itself is called tauberian. In this section we establish a tauberian theorem of Wiener-Ikehara type which will be used to prove the prime number problem.

We begin by introducing an analogue for integrals of the Fejér kernel which occurs in Lemma 8A. For each positive real number λ, we define a function K_λ on the reals by

$$K_\lambda(x) = \frac{1}{2}\int_{-2\lambda}^{2\lambda}\left(1 - \frac{|t|}{2\lambda}\right)e^{ixt}\,dt.$$

Of course, K_λ is not related to the function K_σ of Section 9. We list some important properties of K_λ in the following lemma.

LEMMA 10A: *Let x be real, $\lambda > 0$ and $\delta > 0$. Then*

$$K_\lambda(x) = \lambda\left(\frac{\sin \lambda x}{\lambda x}\right)^2, \tag{10.1}$$

$$0 \leqq K_\lambda(x) \leqq \min(\lambda, \lambda^{-1}x^{-2}), \tag{10.2}$$

$$\int_{-\infty}^{\infty} K_\lambda(x)\,dx \tag{10.3}$$

exists and is a positive number independent of λ,

$$\int_{|u|>\delta} K_\lambda(u)\,du \leqq \frac{2}{\lambda\delta}. \tag{10.4}$$

REMARKS: *For sufficiently large λ, the function $t \mapsto 1 - |t|/(2\lambda)$ is close to the constant function* 1 *on a fixed bounded set, and, moreover, the main contribution to $\int_{-\infty}^{\infty} K_\lambda(x)\,dx$ occurs in a small neighborhood of the origin. The last assertion follows from* (10.4) *by taking λ to be a number which is large relative to $1/\delta$. This is a special case of a well known principle of harmonic analysis.*

We do not need to find the value of the integral in (10.3). However, it is well known that

$$\int_{-\infty}^{\infty} K_\lambda(x)\,dx = \pi.$$

Proof: $K_\lambda(0) = \lambda$ by inspection. For $x \neq 0$ we write

$$K_\lambda(x) = \int_0^{2\lambda} \left(1 - \frac{t}{2\lambda}\right) \cos xt \, dt$$

and integrate by parts obtaining

$$K_\lambda(x) = \frac{1 - \cos 2\lambda x}{2\lambda x^2}.$$

The proof of (10.1) is completed by using a familiar trigonometric identity.

We prove (10.2) by noting that

$$0 \leqq u^{-2} \sin^2 u \leqq \min (1, u^{-2}).$$

The integral in (10.3) exists by (10.2), and by a change of variable is seen to be equal to $\int_{-\infty}^{\infty} u^{-2} \sin^2 u \, du$, which is independent of λ.

We use (10.2) to show (10.4):

$$\int_{|u|>\delta} K_\lambda(u) \, du \leqq 2 \int_\delta^\infty \lambda^{-1} u^{-2} \, du = \frac{2}{\lambda\delta}. \quad \blacksquare$$

We require another important result from harmonic analysis, the so-called Riemann-Lebesgue lemma. The form in which we ultimately cast the result (Lemma 10C) is a little unusual. It results from our wish to avoid using Lebesgue integration and the completeness of L^1, the space of absolutely integrable functions on $(-\infty, \infty)$. We begin by establishing the Riemann-Lebesgue lemma for a very simple class of functions.

LEMMA 10B: *Let f be a continuous complex valued function on $(-\infty, \infty)$ which is zero off a bounded set. Then*

$$\lim_{y \to \pm\infty} \int_{-\infty}^{\infty} f(t) e^{ity} \, dt = 0.$$

Proof: Let

$$I_y = \int_{-\infty}^{\infty} f(t) e^{ity} \, dt.$$

Then

$$2I_y = \int_{-\infty}^{\infty} f(t)e^{ity}\,dt + \int_{-\infty}^{\infty} f\left(t + \frac{\pi}{y}\right) e^{i\{t+(\pi/y)\}y}\,dt$$

$$= \int_{-\infty}^{\infty} \left\{f(t) - f\left(t + \frac{\pi}{y}\right)\right\} e^{ity}\,dt.$$

By uniform continuity, the expression in braces tends to zero uniformly as $y \to \pm\infty$. Also, the last integrand vanishes outside a fixed bounded set if $|y| \geqq 1$. Thus $2I_y \to 0$ as $y \to \pm\infty$. ▌

LEMMA 10C: *For each real number* $\epsilon \in (0, 1]$, *let* f_ϵ *be a complex valued function defined on* $(-\infty, \infty)$ *with the following properties:*

(i) f_ϵ *is continuous and is nonzero only on a bounded set* $E_\epsilon \subset (-\infty, \infty)$.

(ii) $\lim_{\epsilon,\epsilon' \downarrow 0} \int_{-\infty}^{\infty} |f_\epsilon(t) - f_{\epsilon'}(t)|\,dt = 0.$

Then

$$\lim_{y\to\pm\infty} \lim_{\epsilon\downarrow 0} \int_{-\infty}^{\infty} f_\epsilon(t)e^{iyt}\,dt = 0.$$

Proof (a special case of the double limit theorem): Let $\delta > 0$ be given. By (ii) there is a number $\epsilon_0 > 0$ such that

$$\left|\int_{-\infty}^{\infty} f_{\epsilon_0}(t)e^{iyt}\,dt - \int_{-\infty}^{\infty} f_\epsilon(t)e^{iyt}\,dt\right| < \frac{\delta}{2}$$

for $0 < \epsilon \leqq \epsilon_0$ and for all real numbers y. By the Cauchy convergence principle,

$$\lim_{\epsilon\downarrow 0} \int_{-\infty}^{\infty} f_\epsilon(t)e^{iyt}\,dt$$

exists for each number y. With the above δ and ϵ_0 we have for all y

$$\left|\int_{-\infty}^{\infty} f_{\epsilon_0}(t)e^{iyt}\,dt - \lim_{\epsilon\downarrow 0}\int_{-\infty}^{\infty} f_\epsilon(t)e^{iyt}\,dt\right| \leqq \frac{\delta}{2}.$$

Now f_{ϵ_0} is a continuous function which is nonzero only on a bounded set. By the preceding lemma,

$$\left|\int_{-\infty}^{\infty} f_{\epsilon_0}(t)e^{iyt}\,dt\right| < \frac{\delta}{2}$$

provided $|y| > Y = Y(\epsilon_0, \delta)$. By the triangle inequality,

$$|\lim_{\epsilon \downarrow 0} \int_{-\infty}^{\infty} f_\epsilon(t) e^{iyt}\, dt| < \delta$$

provided $|y| > Y$. ∎

With these preparations completed, we turn to our tauberian theorem.

THEOREM 10D: *Suppose F is a nondecreasing real valued function on $[0, \infty)$ such that for each number $\sigma > 1$ we have*

$$\int_0^\infty |F(x)| e^{-\sigma x}\, dx < \infty.$$

Suppose further that

$$(10.5) \qquad \int_0^\infty F(x) e^{-sx}\, dx = \frac{L}{s-1} + G(s) \quad (\mathrm{Re}\ s > 1),$$

where G is a continuous function on Re $s > 1$ *and satisfies*

$$(*) \qquad \lim_{\sigma, \sigma' \downarrow 1} \int_{-\lambda}^{\lambda} |G(\sigma + it) - G(\sigma' + it)|\, dt = 0$$

for every fixed positive number λ. Then

$$\lim_{x \to \infty} e^{-x} F(x) = L.$$

REMARKS: *Consider the condition*

$$(**) \quad \begin{cases} G \text{ can be extended to the closed half plane } \mathrm{Re}\ s \geqq 1 \\ \text{as a continuous function.} \end{cases}$$

If we replace (*) by (**) we have the classical theorem of Wiener and Ikehara (cf. [30]). Since (**) implies (*), the present theorem is a slight generalization of the Wiener-Ikehara theorem. Our proof follows closely that of Landau [14], except for the more cautious application of the Riemann-Lebesgue lemma.

We must assume that (*) holds for *all* positive λ. If we have (*) for each positive $\lambda < \lambda_0$, then we may only deduce that

$$A_1 L \leqq \varliminf_{x \to \infty} e^{-x} F(x) \leqq \varlimsup_{x \to \infty} e^{-x} F(x) \leqq A_2 L,$$

where $A_1 = A_1(\lambda_0)$, $A_2 = A_2(\lambda_0)$, and $0 < A_1 < 1 < A_2$ (cf. [3],

[9], [10]). We illustrate this phenomenon by a simple example. Let λ_0 be a positive number and define a function F on $[0, \infty)$ by

$$F(x) = \int_0^x (1 - \cos \lambda_0 u) e^u \, du.$$

Integration by parts shows that

$$\int_0^\infty e^{-sx} F(x) \, dx = \frac{1}{s-1} - \frac{1}{s} - \frac{s-1}{s\{(s-1)^2 + \lambda_0^2\}} \quad (\operatorname{Re} s > 1).$$

Here the function G has poles at $1 \pm i\lambda_0$ and thus satisfies (*) only if $\lambda < \lambda_0$. On the other hand, the expression defining F may be explicitly integrated giving

$$F(x) = e^x\{1 - (1 + \lambda_0^2)^{-1/2} \cos (\lambda_0 x - \arctan \lambda_0)\} - \lambda_0^2 (1 + \lambda_0^2)^{-1}.$$

We therefore have

$$\overline{\lim_{x \to \infty}}\, e^{-x} F(x) = 1 + (1 + \lambda_0^2)^{-1/2}$$

and

$$\underline{\lim_{x \to \infty}}\, e^{-x} F(x) = 1 - (1 + \lambda_0^2)^{-1/2}.$$

Proof of Theorem 10D. We may assume $F(0) \geqq 0$. If $F(0) < 0$, we could take $F_1 = F + c$ (where $c \geqq -F(0)$), define G_1 by $G_1(s) = G(s) + c/s$, and carry out our argument for F_1 and G_1 in place of F and G.

We have

$$G(s) = \int_0^\infty \{F(x) - L\, e^x\} e^{-sx} \, dx.$$

Set $s = 1 + \epsilon + it$, $\epsilon > 0$, multiply each side of the last equation by $\frac{1}{2}\left(1 - \frac{|t|}{2\lambda}\right) e^{ity}$, and integrate. We obtain

$$\begin{aligned} &\frac{1}{2} \int_{-2\lambda}^{2\lambda} \left(1 - \frac{|t|}{2\lambda}\right) e^{iyt} G(1 + \epsilon + it) \, dt \\ &\quad = \frac{1}{2} \int_{-2\lambda}^{2\lambda} \left(1 - \frac{|t|}{2\lambda}\right) e^{ity} \left\{ \int_0^\infty (F(x) - Le^x) e^{-x(1+\epsilon+it)} \, dx \right\} dt \\ &\quad = \int_0^\infty e^{-(1+\epsilon)x} (F(x) - Le^x) \left\{ \frac{1}{2} \int_{-2\lambda}^{2\lambda} \left(1 - \frac{|t|}{2\lambda}\right) e^{it(y-x)} \, dt \right\} dx. \end{aligned}$$

The interchange of integrations is justified by the convergence of $\int_0^\infty |F(x) - Le^x|e^{-x(1+\epsilon)}\,dx$ (cf. Appendix 3G). In terms of the function K_λ, we have

$$(10.6)\quad \frac{1}{2}\int_{-2\lambda}^{2\lambda}\left(1 - \frac{|t|}{2\lambda}\right)G(1 + \epsilon + it)e^{iyt}\,dt$$
$$= \int_0^\infty e^{-(1+\epsilon)x}F(x)K_\lambda(y - x)\,dx - L\int_0^\infty e^{-\epsilon x}K_\lambda(y - x)\,dx.$$

Now let $\epsilon \downarrow 0$ in (10.6). Since $K_\lambda \geqq 0$, we have by Appendix 5 and a change of variable

$$\lim_{\epsilon\downarrow 0} L\int_0^\infty e^{-\epsilon x}K_\lambda(y - x)\,dx = L\int_0^\infty K_\lambda(y - x)\,dx = L\int_{-\infty}^{y} K_\lambda(u)\,du.$$

The last integral is finite by Lemma 10A.

Let $\mathcal{G}_\epsilon(y)$ denote the left side of (10.6). We have

$$|\mathcal{G}_\epsilon(y) - \mathcal{G}_{\epsilon'}(y)| \leqq \frac{1}{2}\int_{-2\lambda}^{2\lambda} |G(1 + \epsilon + it) - G(1 + \epsilon' + it)|\,dt \to 0$$

as ϵ, $\epsilon' \downarrow 0$. By the Cauchy convergence principle, $\lim_{\epsilon\downarrow 0} \mathcal{G}_\epsilon(y)$ exists (as a finite limit).

Using Appendix 5 again, we have

$$\lim_{\epsilon\downarrow 0} \int_0^\infty e^{-\epsilon x}e^{-x}F(x)K_\lambda(y - x)\,dx = \int_0^\infty e^{-x}F(x)K_\lambda(y - x)\,dx.$$

We therefore deduce from (10.6) that

$$\lim_{\epsilon\downarrow 0} \mathcal{G}_\epsilon(y) = \int_0^\infty e^{-x}F(x)K_\lambda(y - x)\,dx - L\int_{-\infty}^{y} K_\lambda(u)\,du.$$

Since two terms in the above expression are finite, we see that the integral involving F is finite for each y.

Next, we let $y \to \infty$. We apply our version of the Riemann-Lebesgue lemma (Lemma 10C) with

$$f_\epsilon(t) = \begin{cases} \frac{1}{2}\left(1 - \frac{|t|}{2\lambda}\right)G(1 + \epsilon + it), & |t| \leqq 2\lambda \\ 0, & |t| > 2\lambda \end{cases}$$

and obtain

$$\lim_{y\to\infty}\lim_{\epsilon\downarrow 0} \mathcal{G}_\epsilon(y) = 0.$$

Also, Lemma 10A shows that (as $y \to \infty$)

$$\int_{-\infty}^{y} K_\lambda(u)\,du = \int_{-\infty}^{\infty} K_\lambda(u)\,du + o(1).$$

For each fixed $\lambda > 0$ we therefore have (as $y \to \infty$)

$$\int_0^\infty e^{-x}F(x)K_\lambda(y-x)\,dx = L\int_{-\infty}^{\infty} K_\lambda(u)\,du + o(1).$$

We set

$$K_\lambda^*(t) = \frac{K_\lambda(t)}{\int_{-\infty}^{\infty} K_1(u)\,du}.$$

In view of (10.3) K_λ^* has total integral 1. We then obtain

$$\int_0^\infty e^{-x}F(x)K_\lambda^*(y-x)\,dx = L + o(1). \tag{10.7}$$

The function of y implied by the o-symbol depends on λ. We see from Equation (10.7) that $L \geqq 0$.

Setting $\delta = \left(\frac{1}{2}\lambda\int_{-\infty}^{\infty} K_1(u)\,du\right)^{-1/2}$, we have by (10.4)

$$\int_{|u|>\delta} K_\lambda^*(u)\,du \leqq \frac{2}{\lambda\delta\int_{-\infty}^{\infty} K_1(u)\,du} = \delta. \tag{10.8}$$

Let $J(y, \lambda)$ denote the left side of (10.7). Since the integrand in (10.7) is nonnegative and F is monotone, for each $y \geqq \delta$ we have

$$\begin{aligned} J(y,\lambda) &\geqq \int_{y-\delta}^{y+\delta} \geqq F(y-\delta)e^{-y-\delta}\int_{-\delta}^{\delta} K_\lambda^*(u)\,du \\ &\geqq F(y-\delta)e^{-(y-\delta)}e^{-2\delta}(1-\delta). \end{aligned} \tag{10.9}$$

Let $\epsilon > 0$ be given. We may take λ large enough so that $e^{2\delta}(1-\delta)^{-1} < 1 + \epsilon$. With this choice of λ we have by Equations (10.7) and (10.9)

$$F(y-\delta)e^{-(y-\delta)} \leqq L(1+\epsilon) + o(1)$$

or

$$\overline{\lim_{y\to\infty}}\, e^{-y}F(y) \leqq L(1+\epsilon).$$

Since this relation holds for each $\epsilon > 0$,

$$\overline{\lim_{y\to\infty}}\, e^{-y}F(y) \leqq L. \tag{10.10}$$

We now obtain the inequality in the opposite direction. Since $e^{-y}F(y)$ is bounded, there exists a positive number B such that for all $y \geqq 0$ we have $e^{-y}F(y) \leqq B$. Now for all positive λ and y we have by the monotonicity of F and by (10.8)

$$\begin{aligned} J(y, \lambda) &\leqq B \int_{|u|>\delta} K_\lambda^*(u)\, du + F(y+\delta)e^{-y+\delta} \int_{-\delta}^{\delta} K_\lambda^*(u)\, du \\ &\leqq B\delta + F(y+\delta)e^{-(y+\delta)}e^{2\delta} \int_{-\infty}^{\infty} K_\lambda^*(u)\, du \\ &\leqq \{B\delta + F(y+\delta)e^{-(y+\delta)}\}e^{2\delta}. \end{aligned}$$

The last inequality and (10.7) imply that (as $y \to \infty$)

$$F(y)e^{-y} \geqq Le^{-2\delta} - B\delta + o(1)$$

for each fixed pair λ, δ with $\lambda\delta^2 = \left(\frac{1}{2}\int_{-\infty}^{\infty} K_1(u)\, du\right)^{-1}$.
Thus, for each positive δ we have

$$\varliminf_{y\to\infty} F(y)e^{-y} \geqq Le^{-2\delta} - B\delta$$

and so

$$\varliminf_{y\to\infty} F(y)e^{-y} \geqq L.$$

Finally, by (10.10) and the last inequality, we have

$$\lim_{y\to\infty} F(y)e^{-y} = L. \blacksquare$$

11. PROOF OF THE PRIME NUMBER THEOREM

We now assemble the various results needed to prove Theorem 7A. Under the hypotheses of the theorem, we showed in Section 8 that ζ_P has no zeros on $\operatorname{Re} s = 1$. In Section 9 we proved that

$$\lim_{\sigma,\sigma' \downarrow 1} \int_{-\lambda}^{\lambda} |h(\sigma + it) - h(\sigma' + it)|^2\, dt = 0$$

for each fixed positive number λ, where

$$h(s) = -\frac{1}{s}\frac{\zeta'(s)}{\zeta(s)} - \frac{1}{s-1}.$$

By Schwarz's inequality, we have

$$\lim_{\sigma,\sigma' \downarrow 1} \int_{-\lambda}^{\lambda} |h(\sigma + it) - h(\sigma' + it)|\, dt = 0$$

for each fixed positive number λ. We express $-\frac{1}{s}\frac{\zeta'(s)}{\zeta(s)}$ as a Laplace integral, using (2.9) and a change of variable:

$$-\frac{1}{s}\frac{\zeta'(s)}{\zeta(s)} = \int_1^\infty x^{-s-1}\psi(x)\, dx = \int_0^\infty e^{-su}\psi(e^u)\, du.$$

In the tauberian theorem of the preceding section, we take

$$L = 1, \quad F(x) = \psi(e^x), \quad \text{and} \quad G(s) = h(s)$$

and conclude that

$$\lim_{x\to\infty} e^{-x}\psi(e^x) = 1.$$

Lemma 2E now tells us that the P.N.T. holds. ▮

12. OTHER HYPOTHESES, OTHER CONCLUSIONS

In this section we shall discuss briefly, and without proof, the following related problems.

A. The connection between the prime number theorem and the absence of zeros of zeta on Re $s = 1$.

B. The remainder estimates which can be made for $\pi(x) - \int_2^x (\log u)^{-1}\, du$.

C. Results that can be obtained by use of methods that are elementary in the sense of Selberg.

D. The converse of the prime number problem: If $\pi_P(x)$ is close to $\int_2^x (\log u)^{-1}\, du$, is $I_P(x)$ then close to Ax for some constant $A = A_P$?

E. The Riemann hypothesis for g-numbers.

A. The nonvanishing of the Riemann zeta function on Re $s = 1$ was an important element in the above proof of the prime number theorem. Conversely, a well known theorem asserts that the truth of the P.N.T. (for ordinary numbers) implies the nonvanishing of zeta for Re $s \geqq 1$ (*cf.* [10]). A similar situation applies for g-numbers. Specifically, we have the following theorem [7].

Let P be a g-prime system for which the P.N.T. is true and which satisfies (7.1) *with $\gamma > 1$. Let ζ_P be the associated zeta function which has been extended to* Re $s \geqq 1$, $s \neq 1$ (*cf. Lemma* 7B). *Then for each real number $t \neq 0$, $\zeta(1 + it) \neq 0$.*

This result may be proved by the argument used in Theorem 8E, except that the estimate given in Lemma 8C must be replaced. By using the P.N.T. we may show that for each $\epsilon > 0$ and $t \neq 0$ there is a $C = C_P(\epsilon, t) > 0$ such that

$$|\zeta(\sigma + it)| > C(\sigma - 1)^{\epsilon} \quad (1 < \sigma < 2).$$

The theorem follows from this estimate and Lemma 8D.

B. Let $R = R_P$ (the "remainder function") be defined on (p_1, ∞) by

$$R(x) = \pi_P(x) - \int_{p_1}^{x} \log^{-1} u \, du.$$

Theorem 7A asserts that if (7.1) holds with $\gamma = \frac{3}{2} + \epsilon$, then $R(x) = o(x \log^{-1} x)$. If P is a g-prime system for which the number γ is large, we might hope for a better estimate for R than that given above.

It was proved by Nyman [20] that *if* (7.1) *holds for arbitrarily large positive numbers γ, then*

$$R(x) = O(x \log^{-\delta} x) \tag{12.1}$$

for arbitrarily large positive values of δ. The constant implied by the O depends on δ. Nyman's proof involved estimates of high order derivatives of ζ'/ζ and used an application of Plancherel's theorem in its tauberian argument.

Malliavin [18] attacked this problem using the same strategy as Nyman, but altered his tactics so as to be able to estimate the constant implied by the O in (12.1). This approach enabled him to handle finer error terms, and he proved that *if $\alpha \in (0, 1)$ and $c_1 > 0$ and*

$$I(x) = Ax + O\{x \exp(-c_1 \log^{\alpha} x)\} \tag{12.2}$$

then

$$R(x) = O\{x \exp(-c_2 \log^\beta x)\}, \tag{12.3}$$

where $c_2 > 0$ *and* $\beta = \alpha/10$. Hall [7] improved Malliavin's result by showing that (12.2) *implies* (12.3) *with* $\beta = \alpha/8$.

C. In his paper giving an elementary proof of the prime number theorem, A. Selberg [25] observed that his argument could be applied to g-numbers, but required slightly stronger hypotheses than Beurling had made. This was spelled out explicitly by Amitsur [1], who gave an elementary proof of the P.N.T. in the case that (7.1) holds with $\gamma > 2$. Error estimates of the type (12.1) with the value of δ dependent on γ were obtained by Wegmann [29] using methods derived from an elementary proof of the P.N.T. given by Wirsing [31].

D. In [20], Nyman also considered the converse of the prime number problem. He proved that *if* (12.1) *holds for arbitrarily large positive numbers* δ, *then* (7.1) *holds for some number* $A = A_P$ *and arbitrarily large positive numbers* γ. Malliavin [18] showed that *if* (12.3) *holds for* $c_2 > 0$ *and* $\beta \in (0, 1)$, *then* (12.2) *holds for some numbers* $A > 0$, $c_1 > 0$, *and* $\alpha = \beta/(2 + \beta)$. Nyman and Malliavin obtained these results using Fourier analysis and tauberian methods. Malliavin observed, however, that the problem of estimating $I(x)$ from an assumption on the asymptotic behavior of the function π seemed to be of an abelian character (in the sense that the function I can be obtained from π by a sort of averaging process). Diamond [5] gave an abelian proof of this result using methods from the operational calculus. He improved Malliavin's result by showing that *if* (12.3) *holds for* $c_2 = 1$ *and* $\beta \in (0, 1)$, *then*

$$I(x) = Ax + O(x \exp\{-(\log x \log\log x)^\alpha\}),$$

for some number $A = A_P$ *and* $\alpha = \beta/(1 + \beta)$.

E. The most famous unsolved problem in prime number theory is the Riemann hypothesis, which may be stated in the following form: if $\zeta(s) = 0$, then Re $s \leqq \frac{1}{2}$. Here ζ is the Riemann zeta function of the ordinary prime numbers. An equivalent form of the Riemann hypothesis is the assertion

$$R(x) = O(x^{(1/2)+\epsilon}) \tag{12.4}$$

for each $\epsilon > 0$ [17; Satz 453].

Suppose P is a g-prime system for which

$$I_P(x) = Ax + O(x^\theta) \tag{1.4$'$}$$

for some $\theta \leqq \frac{1}{2}$. In this case it is possible to continue ζ_P analytically to the set Re $s > \theta$. The Riemann hypothesis for such a system P is the assertion that ζ_P has no zeros in Re $s > \frac{1}{2}$. This assertion is not generally true for g-numbers, as Malliavin [18] showed by a continuous analogue of a g-prime system.

In the Introduction, we stated that Landau proved that if (1.4′) holds for some $\theta < 1$, then there is a positive constant c such that

$$R_P(x) = O\{x \exp(-c\sqrt{\log x})\}. \tag{12.5}$$

(This is the error term which occurs in de la Vallée Poussin's proof of the P.N.T.) Except perhaps for the value of the constant c, Landau's theorem is essentially best possible for generalized numbers. This may be seen from an example of Hall [7] in which (1.4′) holds but

$$|R_P(x)| > Bx \exp(-c\sqrt{\log x})$$

for positive numbers B and c and a sequence of numbers x tending to infinity. The device used by Hall was the creation of a zeta function with zeros at points s_n (and hence, by reflection, at $\overline{s_n}$) where Re $s_n \uparrow 1$ and Im $s_n \to \infty$ as $n \to \infty$ and the sum of the orders of the zeros was suitably small.

The above examples show that a successful proof of the Riemann hypothesis will require more than the facts that the integers are a multiplicative semigroup generated by the primes and $[x]$ is close to x.

13. SOME CONJECTURES AND UNSOLVED PROBLEMS

To conclude this survey, we list some unsolved problems on the asymptotic distribution of generalized numbers.

A. We proved the P.N.T. for generalized numbers in the case that

$$(13.1)\quad x^{-1}\{I_P(x) - Ax\} \log x = O(\{\log x\}^{1-\gamma}) \qquad (\gamma > \tfrac{3}{2}).$$

If the weaker condition

$$(13.2)\qquad \int_1^\infty (x^{-1}\{I_P(x) - Ax\} \log x)^2 x^{-1}\, dx < \infty$$

holds and if ζ_P is known to have no zeros with real part one, the arguments of Sections 9–11 can be applied without change to yield the P.N.T. We conjecture that (13.2) would actually imply the nonexistence of zeros with real part one (and hence the validity of the P.N.T. for the g-prime system P). We remark that an elaboration of the methods which we have used will produce a proof of the P.N.T. under a hypothesis which is intermediate between Beurling's condition (13.1) and the condition (13.2); the hypothesis is that there exists a number $\eta > \frac{1}{3}$ such that

$$x^{-1}\{I_P(x) - Ax\} \log x = O\{(\log x)^{-1/2} \exp(-\{\log \log x\}^\eta)\}.$$

B. The error terms of Malliavin and Hall in (12.3) have index $\beta = \alpha/10$ and $\alpha/8$, respectively. (Actually, Hall obtains the slightly better result $\beta = \alpha/7.91$.) There is quite likely room for improvement here, possibly to the value $\beta = \alpha/(1 + \alpha)$. This last number occurs in the abelian case discussed in 12D, and if $\alpha = 1$, is the index in Hall's example mentioned in 12E.

C. For the moment let P be the sequence of ordinary primes and let ψ_P be the function defined in Section 2. Chebyshev made an important early contribution to prime number theory by proving that

$$.92x < \psi_P(x) < 1.06x$$

for all sufficiently large x(cf. [10], [15], [19], [23]). Even today, the estimate

$$(13.3)\qquad \psi_P(x) = O(x)$$

is useful in many problems (e.g., in proving Selberg's formula, which is the basis of elementary proofs of the P.N.T.). We ask when (13.3) holds for generalized numbers. On one hand, (13.3) is certainly valid for systems P for which (7.1) holds with $\gamma > \frac{3}{2}$. On the other hand, (13.3) is not generally true for systems P for which $\gamma < 1$, as an example of Hall [7] shows. We would guess that (13.3) is valid for any g-prime system P for which (7.1) holds with $\gamma > 1$.

D. We stated in 12C that Beurling's theorem could be established by elementary methods if (7.1) holds with $\gamma > 2$. No one has yet succeeded in using elementary methods to establish the P.N.T. or even Chebyshev's estimate (13.3) in the case $\frac{3}{2} < \gamma < 2$.

E. Let I_P denote, as before, the integer counting function of some g-prime system P. Beurling has conjectured that if

$$|I_P(x) - [x]| < E(x) \quad (1 \leqq x < \infty)$$

for a suitable estimation function E, then P must in fact be the ordinary primes and $I_P(x) = [x]$ for all $x \geqq 1$. Since I_P and $[\ \ ]$ are integer valued, the conjecture is certainly true for $E = 1$. For what larger estimation functions is Beurling's conjecture true? The following example shows that E must be a rather "small" function. Let P' be the ordinary primes with 2 and 3 deleted and $\frac{3}{2}$ inserted as a prime. It can be verified that

$$|I_{P'}(x) - [x]| < 1 + \frac{\frac{2}{3}}{\log \frac{3}{2}} \log x.$$

Thus

$$E(x) = 1 + 2 \log x$$

is too large to allow the conjecture to be true.

Appendices

Appendix 1. O and o.

Let g be a positive valued function defined on some interval of the form $[c, \infty)$. In this paper the expression $O(g(x))$ denotes a complex valued function G defined on some interval $[c', \infty)$ with the following property: there exist constants $K > 0$ and $x_0 \geqq \max(c, c')$ such that

$$|G(x)| \leqq Kg(x)$$

for all $x \geqq x_0$. The expression $o(g(x))$ denotes a complex valued function G defined on some interval $[c', \infty)$ such that

$$\lim_{x \to \infty} \frac{G(x)}{g(x)} = 0.$$

In particular,

$$f(x) = g(x) + o(g(x)) \text{ iff } \lim_{x \to \infty} f(x)/g(x) = 1,$$

and

$$f(x) = L + o(1) \text{ iff } \lim_{x \to \infty} f(x) = L.$$

Appendix 2. RIEMANN-STIELTJES INTEGRALS

We shall define a Riemann-Stieltjes (R.S.) integral of a special type which suffices for our purposes and shall justify integration by parts for this integral. For a more complete discussion of R.S. integrals see, for example, [2].

Let f be a continuous function on $[1, \infty)$ and let $1 = x_1 < x_2 < \cdots$ be a sequence of points which may be finite or infinite in number. In the latter case we assume that there are only a finite number of points x_i in each interval of the form $[1, N]$. Let g be a function defined on $[1, \infty)$ which is constant on each interval $[x_{i-1}, x_i)$, $i \geqq 2$. For finite x, we set

$$\int_{1-}^{x} f(t)\, dg(t) = \sum_{x_i \leqq x} f(x_i)\, \Delta g(x_i),$$

where for $i \geqq 1$

$$\Delta g(x_i) = g(x_i) - g(x_{i-1})$$

and $g(x_0)$ is defined to be zero. We shall define

$$\int_{1-}^{\infty} f(t)\, dg(t) = \lim_{x \to \infty} \int_{1-}^{x} f(t)\, dg(t)$$

in case this limit exists.

If $g(1) = 0$, we shall require only that f be continuous on $(1, \infty)$. In this case we set $f(x_1)\, \Delta g(x_1) = 0$ (even if $f(1)$ is not defined) and write $\int_1^x$ in place of $\int_{1-}^x$. An example in which f is not continuous at 1 is provided by the integral $\int_1^x (\log t)^{-1}\, d\psi(t)$ occurring in the proof of Lemma 2E.

THEOREM: (integration by parts for special R.S. integrals.) *Let g satisfy the requirements of the above definition and let f be continuous and have a piecewise continuous derivative on* $[1, \infty)$ $\{$*on* $(1, \infty)$ *in the case* $g(1) = 0\}$. *If* $x \geqq 1$, *then*

$$\int_{1-}^x f(t)\, dg(t) = f(x)g(x) - \int_1^x g(t)f'(t)\, dt,$$

where the integral on the right is a Riemann integral.

Proof: First we treat the case in which f is continuous at 1. Suppose $x_I \leqq x < x_{I+1}$, where g is constant on $[x_I, x_{I+1})$. Then

$$\begin{aligned}
\int_{1-}^x f\, dg &= \sum_{i=1}^{I} f(x_i)\, \Delta g(x_i) \\
&= \sum_{i=1}^{I} f(x_i)g(x_i) - \sum_{i=0}^{I-1} f(x_{i+1})g(x_i) \\
&= f(x_I)g(x_I) - \sum_{i=1}^{I-1} g(x_i)\{f(x_{i+1}) - f(x_i)\} \\
&= f(x)g(x) - g(x_I)\{f(x) - f(x_I)\} \\
&\qquad\qquad - \sum_{i=1}^{I-1} g(x_i)\{f(x_{i+1}) - f(x_i)\} \\
&= f(x)g(x) - \int_{x_I}^x g(t)f'(t)\, dt - \sum_{i=1}^{I-1} \int_{x_i}^{x_{i+1}} g(t)f'(t)\, dt \\
&= f(x)g(x) - \int_1^x g(t)f'(t)\, dt.
\end{aligned}$$

If $g(1) = 0$ and f is continuous and piecewise continuously differentiable only on $(1, \infty)$, we shall alter f so as to be able to use the above reasoning. Define a new function F on $[1, \infty)$ by

$$F(x) = \begin{cases} f(x), & x_2 \leqq x < \infty, \\ f(x_2), & 1 \leqq x < x_2. \end{cases}$$

Since g is zero on $[1, x_2)$, we have

$$\int_{1-}^x f\, dg = \int_{1-}^x F\, dg$$

and

$$F(x)g(x) - \int_1^x F'(t)g(t)\,dt = f(x)g(x) - \int_1^x f'(t)g(t)\,dt.$$

The theorem is true for F, and by the above equalities, the theorem also holds for f.

Appendix 3. SOME THEOREMS ON INFINITE PROCESSES

We list here the theorems on exchanging operations such as limits, infinite summation, integration, and differentiation which we have used in this article.

A. *Suppose $f_1, f_2, \cdots$ are continuous complex valued functions on a subset S of the complex plane and suppose $\sum_{n=1}^{\infty} f_n$ converges uniformly on S. Then the sum of the series is continuous on S.*

B. *Suppose $f_1, f_2, \cdots$ are continuous complex valued functions on a finite line segment Γ in the complex plane and suppose $\sum_{n=1}^{\infty} f_n$ converges uniformly on Γ. Then*

$$\int_\Gamma \sum_{n=1}^{\infty} f_n(z)\,dz = \sum_{n=1}^{\infty} \int_\Gamma f_n(z)\,dz.$$

C. *Suppose $f_1, f_2, \cdots$ are complex valued functions with continuous derivatives on an open connected set Ω in the complex plane and suppose $\sum_{n=1}^{\infty} f_n$ converges at some point of Ω and $\sum_{n=1}^{\infty} f_n'$ converges uniformly on Ω. Then at each point of Ω we have*

$$\left(\sum_{n=1}^{\infty} f_n\right)' = \sum_{n=1}^{\infty} f_n'.$$

This result can be deduced from B and the fundamental theorem of calculus.

D. *Suppose f is a continuous complex valued function on the bounded rectangle $[a, b] \times [c, d]$ and suppose ϕ is a complex valued Riemann integrable function on the interval $[a, b]$. Then*

$$\int_a^b f(x, y)\phi(x)\, dx$$

is a continuous function of y *for* $c \leqq y \leqq d$.

This is an immediate consequence of the uniform continuity of f.

E. *Suppose* f *is a continuous complex valued function on the bounded rectangle* $[a, b] \times [c, d]$ *and suppose* ϕ *and* ψ *are complex valued Riemann integrable functions on the intervals* $[a, b]$ *and* $[c, d]$, *respectively. Then*

$$(*) \quad \int_a^b \left\{ \int_c^d f(x, y)\psi(y)\, dy \right\} \phi(x)\, dx = \int_c^d \left\{ \int_a^b f(x, y)\phi(x)\, dx \right\} \psi(y)\, dy.$$

This result can be proved by subdividing $[a, b] \times [c, d]$ by horizontal and vertical lines into smaller rectangles. If the diameter of the largest subrectangle is sufficiently small, we can approximate f as closely as desired by a function which is constant on the interior of each subrectangle. For such a step function the assertion of the theorem is immediate.

F. *Suppose* f *is a bounded continuous complex valued function on the half strip* $[a, \infty) \times [c, d]$, *suppose* ϕ *is a complex valued function on* $[a, \infty)$ *which is Riemann integrable on every interval of the form* $[a, b]$, *where* $a < b < \infty$, *and suppose* $\int_a^\infty |\phi(x)|\, dx < \infty$. *Then* $\int_a^\infty f(x, y)\phi(x)\, dx$ *exists and is a continuous function of* y *for* $c \leqq y \leqq d$.

This result follows from D by a simple truncation argument.

G. *Suppose* f *is a bounded continuous complex valued function on the half strip* $[a, \infty) \times [c, d]$, *suppose* ψ *is a complex valued Riemann integrable function on* $[c, d]$, *suppose* ϕ *is a complex valued function on* $[a, \infty)$ *which is Riemann integrable on every interval of the form* $[a, b]$, *where* $a < b < \infty$, *and suppose* $\int_a^\infty |\phi(x)|\, dx < \infty$. *Then*

$$(**)\quad \int_a^\infty \left\{\int_c^d f(x, y)\psi(y)\, dy\right\} \phi(x)\, dx = \int_c^d \left\{\int_a^\infty f(x, y)\phi(x)\, dx\right\} \psi(y)\, dy.$$

This result is used in an essential way at the beginning of the proof of Theorem 10D. It may be deduced from E as follows. Given $\epsilon > 0$, we choose b so large that $\int_b^\infty |\phi(x)|\, dx \leqq \epsilon/M$, where

$$M = \int_c^d |\psi(y)|\, dy \cdot \sup |f(x, y)|.$$

In view of E, we have (*). But by our choice of b the left-hand side of (*) differs from the left-hand side of (**) by at most ϵ, and similarly for the right-hand sides.

H. *Let Ω be an open set in the complex plane. Suppose g and f are bounded continuous complex valued functions on $[a, \infty) \times \Omega$ such that $\frac{\partial}{\partial s} g(x, s) = f(x, s)$. Suppose ϕ is a complex valued function on $[a, \infty)$ which is Riemann integrable on every interval of the form $[a, b]$, where $a < b < \infty$, and suppose $\int_a^\infty |\phi(x)|\, dx < \infty$. Then at each point s of Ω we have*

$$\frac{d}{ds} \int_a^\infty g(x, s)\phi(x)\, dx = \int_a^\infty f(x, s)\phi(x)\, dx.$$

This result can be deduced from G and the fundamental theorem of calculus. It can be used to justify the differentiation under the integral sign carried out at the beginning of Section 9. To see this, we take

$$\Omega = \{s\colon \operatorname{Re} s > 1 + 2\epsilon,\ |s| < \epsilon^{-1}\}, \quad \phi(x) = x^{-2-\epsilon}\{I(x) - Ax\},$$
$$g(x, s) = sx^{-s+1+\epsilon}, \quad f(x, s) = x^{-s+1+\epsilon}(1 - s \log x).$$

Appendix 4. BESSEL INEQUALITY FOR FOURIER TRANSFORMS

THEOREM: *Let f be a complex valued function defined on the reals which is Riemann integrable on each interval of the form $[-M, M]$ and suppose*

$\int_{-\infty}^{\infty} |f(t)|\, dt < \infty$. *Let*

$$F(x) = \int_{-\infty}^{\infty} f(t)e^{-ixt}\, dt.$$

If λ *is any positive number, then*

$$\int_{-\lambda}^{\lambda} |F(x)|^2\, dx \leqq 2\pi \int_{-\infty}^{\infty} |f(t)|^2\, dt.$$

REMARK: *It is clear that we may let* $\lambda \to \infty$ *in the last formula. In this case a theorem of Plancherel asserts that actually*

$$\int_{-\infty}^{\infty} |F(x)|^2\, dx = 2\pi \int_{-\infty}^{\infty} |f(t)|^2\, dt.$$

Proof: Assume first that f is zero off $[-M, M]$, where M is a positive integer. For positive integral k, we approximate F by

$$F_k(x) = \int_{-M}^{M} f(t)e^{-i[kt]x/k}\, dt.$$

For α and β real,

$$|e^{i\alpha} - e^{i\beta}| = |e^{i(\alpha-\beta)/2} - e^{i(\beta-\alpha)/2}| = 2 \sin\left|\frac{\alpha - \beta}{2}\right| \leqq |\alpha - \beta|,$$

and so

$$|e^{-itx} - e^{-i[kt]x/k}| \leqq \frac{|x|(kt - [kt])}{k} \leqq \frac{|x|}{k}.$$

Thus we have

$$|F(x) - F_k(x)| \leqq \int_{-M}^{M} |f(t)||e^{-itx} - e^{-i[kt]x/k}|\, dt$$
$$\leqq \frac{|x|}{k} \int_{-M}^{M} |f(t)|\, dt,$$

and $F_k \to F$ uniformly on $[-\lambda, \lambda]$ as $k \to \infty$.

We write F_k as a sum

$$F_k(x) = \sum_{n=-Mk}^{Mk-1} a_n e^{-inx/k},$$

where

$$a_n = \int_{n/k}^{(n+1)/k} f(t)\, dt \quad (-Mk \leqq n \leqq Mk - 1).$$

By Schwarz's inequality

$$|a_n|^2 \leqq k^{-1} \int_{n/k}^{(n+1)/k} |f(t)|^2 \, dt,$$

and thus

$$(*) \qquad k \sum_{n=-Mk}^{Mk-1} |a_n|^2 \leqq \sum_{n=-Mk}^{Mk-1} \int_{n/k}^{(n+1)/k} |f|^2 = \int_{-M}^{M} |f|^2.$$

We use the sum for F_k and the orthogonality of exponentials to obtain

$$\int_{-k\pi}^{k\pi} F_k(x)\overline{F_k(x)} \, dx = \sum_{m,n} a_m \overline{a_n} \int_{-k\pi}^{k\pi} e^{i(n-m)x/k} \, dx = 2\pi k \sum_{n=-Mk}^{Mk-1} |a_n|^2.$$

For $k\pi \geqq \lambda$, the preceding equality and (*) imply that

$$\int_{-\lambda}^{\lambda} |F_k(x)|^2 \, dx \leqq 2\pi k \sum_{n=-Mk}^{Mk-1} |a_n|^2 \leqq 2\pi \int_{-M}^{M} |f(t)|^2 \, dt.$$

Now

$$\int_{-\lambda}^{\lambda} |F(x)|^2 \, dx = \lim_{k\to\infty} \int_{-\lambda}^{\lambda} |F_k(x)|^2 \, dx \leqq 2\pi \int_{-\infty}^{\infty} |f(t)|^2 \, dt,$$

and the theorem is proved for functions f which vanish outside the bounded set $[-M, M]$.

We use the special case to prove the general result. Let f be an arbitrary function satisfying the hypotheses of the theorem. For M a positive integer, set

$$F(x; M) = \int_{-M}^{M} f(t) e^{-ixt} \, dt.$$

Since $\int_{-\infty}^{\infty} |f|$ exists, $F(\ ; M) \to F$ uniformly as $M \to \infty$. Thus we have

$$\int_{-\lambda}^{\lambda} |F|^2 = \lim_{M\to\infty} \int_{-\lambda}^{\lambda} |F(\ ; M)|^2 \leqq \lim_{M\to\infty} 2\pi \int_{-M}^{M} |f|^2 = 2\pi \int_{-\infty}^{\infty} |f|^2.$$

Appendix 5. A MONOTONE CONVERGENCE THEOREM

THEOREM: *Suppose f is a nonnegative valued function on $[0, \infty)$ which is Riemann integrable on every finite interval of the form $[0, T]$. Suppose further that*

$\int_0^\infty e^{-\epsilon u} f(u)\,du$ *exists for every positive* ϵ. *Then*

$$\lim_{\epsilon \downarrow 0} \int_0^\infty e^{-\epsilon u} f(u)\,du = \int_0^\infty f(u)\,du$$

in the extended sense. That is, either both sides are infinite or both sides are finite and equal.

Proof: For each $\epsilon > 0$ and $T < \infty$ we have

$$\int_0^T e^{-\epsilon u} f(u)\,du \leqq \int_0^T f(u)\,du \leqq \int_0^\infty f(u)\,du.$$

Letting $T \to \infty$, we obtain

$$\int_0^\infty e^{-\epsilon u} f(u)\,du \leqq \int_0^\infty f(u)\,du.$$

The expression on the left increases as ϵ decreases, and thus has a (possibly infinite) limit as $\epsilon \downarrow 0$. The last inequality implies that

$$(*) \qquad \lim_{\epsilon \downarrow 0} \int_0^\infty e^{-\epsilon u} f(u)\,du \leqq \int_0^\infty f(u)\,du.$$

On the other hand, $e^{-\epsilon u} \to 1$ uniformly for $0 \leqq u \leqq T$ as $\epsilon \downarrow 0$. Thus we have

$$\int_0^T f(u)\,du = \lim_{\epsilon \downarrow 0} \int_0^T e^{-\epsilon u} f(u)\,du \leqq \lim_{\epsilon \downarrow 0} \int_0^\infty e^{-\epsilon u} f(u)\,du.$$

If we let $T \to \infty$ on the left side of the last inequality, we obtain

$$(**) \qquad \int_0^\infty f(u)\,du \leqq \lim_{\epsilon \downarrow 0} \int_0^\infty e^{-\epsilon u} f(u)\,du.$$

Inequalities (*) and (**) imply our assertion.

BIBLIOGRAPHY

1. Amitsur, S. A., "Arithmetic linear transformations and abstract prime number theorems," *Canad. J. Math.*, 13 (1961), 83–109.
2. Apostol, T. M., *Mathematical Analysis*. Addison-Wesley, Reading, Mass., 1957.
3. Beurling, A., "Analyse de la loi asymptotique de la distribution des nombres premiers généralisés, I," *Acta Math.*, 68 (1937), 255–291.

4. Bombieri, E., "Sulle formule di A. Selberg generalizzate per classi di funzioni aritmetiche e le applicazioni al problema del resto nel 'Primzahlsatz'," *Riv. Mat. Univ. Parma,* (2) 3 (1962), 393–440.

5. Diamond, H. G., "Asymptotic distribution of Beurling's generalized integers." (To appear in *Ill. J. Math.*)

6. ———, "A set of generalized numbers showing Beurling's theorem to be sharp." (To appear in *Ill. J. Math.*)

7. Hall, R. S., "Theorems about Beurling's generalized primes and the associated zeta function," Ph.D. Thesis, Univ. of Illinois, 1967.

8. Hardy, G. H., and Wright, E. M., *An Introduction to the Theory of Numbers.* 4th ed., Clarendon Press, Oxford, 1960.

9. Heilbronn, H., and Landau, E., "Bemerkungen zur vorstehenden Arbeit von Herrn Bochner," *Math. Zeit.*, 37 (1933), 10–16.

10. Ingham, A. E., *The Distribution of Prime Numbers.* The University Press, Cambridge, 1932.

11. Kac, M., *Statistical Independence in Probability, Analysis and Number Theory,* Carus Math. Monographs, No. 12, Math. Assoc. of America, 1959, particularly 64–71.

12. Korevaar, J., van Aardenne-Ehrenfest, T., and deBruijn, N. G., "A note on slowly oscillating functions," *Nieuw Arch. Wiskunde* (2) 23 (1949), 77–86.

13. Landau, E., "Neuer Beweis des Primzahlsatzes und Beweis des Primidealsatzes," *Math. Ann.*, 56 (1903), 645–670.

14. ———, "Über den Wienerschen neuen Weg zum Primzahlsatz," *Sitzungsberichte Akad. Berlin* (1932), 514–521. Reprinted in Chelsea edition of [15], 917–924.

15. ———, *Handbuch der Lehre von der Verteilung der Primzahlen.* Teubner, Leipzig, 1909. Reprinted by Chelsea Publishing Co., New York, 1953.

16. ———, "Über die Wurzeln der Zetafunktion," *Math. Zeit.*, 20 (1924), 98–104. Reprinted in Chelsea edition of [15], 909–915.

17. ———, *Vorlesungen über Zahlentheorie* II. Hirzel, Leipzig, 1927.

18. Malliavin, P., "Sur le reste de la loi asymptotique de répartition des nombres premiers généralisés de Beurling," *Acta Math.*, 106 (1961), 281–298.

19. Mathews, G. B., *Theory of Numbers*, I. Deighton, Bell & Co., Cambridge, 1892. Reprinted by Chelsea Publishing Co., New York, 1961.

20. Nyman, B., "A general prime number theorem," *Acta Math.*, 81 (1949), 299–307.

21. Pitt, H. R., *Tauberian Theorems*, Tata Institute Monographs, No. 2, Oxford University Press, London, 1958.

22. Pólya, G., and Szegö, G., *Aufgaben und Lehrsätze aus der Analysis*. Springer, Berlin, 1925.

23. Prachar, K., *Primzahlverteilung*. Springer, Berlin, 1957.

24. Rényi, A., "On the density of certain sequences of integers," *Acad. Serbe Sci. Publ. Inst. Math.*, 8 (1955), 157–162.

25. Selberg, A., "An elementary proof of the prime number theorem," *Ann. of Math.*, (2) 50 (1949), 305–313.

26. Titchmarsh, E. C., *The Theory of the Riemann Zeta-Function*. Clarendon Press, Oxford, 1951.

27. Weber, H., "Über Zahlengruppen in algebraischen Körpern," *Math. Ann.*, 49 (1897), 83–100.

28. ———, *Lehrbuch der Algebra*. 2nd ed., Vol. 2. Vieweg, Braunschweig, 1899.

29. Wegmann, H., "Beiträge zur Zahlentheorie auf freien Halbgruppen, II," *J. Reine Angew. Math.*, 221 (1966), 150–159.

30. Wiener, N., *The Fourier Integral and Certain of its Applications*. The University Press, Cambridge, 1933. Reprinted by Dover Publications, New York, 1959.

31. Wirsing, E., "Elementare Beweise des Primzahlsatzes mit Restglied, II," *J. Reine Angew. Math.*, 214/215 (1964), 1–18.

32. Wunderlich, M. C., "Sieve-generated sequences," *Canad. J. Math.*, 18 (1966), 291–299.

INDEX

abscissa of convergence, 160
algebraic integer, 52
algebraic number field, 52
algorithms:
  greatest common divisor, 122
  Jacobi symbol, 124
  power, 124
approximation theorem, 36
arithmetic formulas, 90
arithmetic relations, 90
Artin's conjecture, 21

Beurling's theorem, 156, 176
birational transformation, 7

companion polynomial, 63
computable function, 81
computable set of numbers, 81
computer hardware, software, 120
congruence conditions, 28
continued fractions, 138

Davis normal form, 93
Davis-Putnam-Robinson theorem, 103
decision problem for diophantine equations, 76
degenerate polynomial, 38
diagonal form, 47
diophantine binary relation, 96
diophantine equations, 4, 25
diophantine problems, 26
diophantine set of numbers, 96
equivalent polynomials, 37
equivalent valuations, 36
exponential diophantine equation, 98
exponential diophantine set, 98
exponent of growth, 167

factorization by trial division, 137
Féjer kernel, 178
finite basis for a curve, 12

generalized integers, 155
generalized prime number system, 154
  Riemann hypothesis for, 197
genus of a curve, 7
Gödel-Davis theorem, 91

Hasse principle, 15, 20
Hensel-Newton method, 29, 42, 43, 44
Hilbert map, 42
Hilbert's problem, 76, 107

infinite descent, method of, 15–16

Kleene normal form, 87

list, 81
listable set, 81
local fields, 37

method, 81
Meyer's theorem, 20
multiprecise arithmetic, 135

O, *o* notation, 200
order of a *p*-adic integer, 31, 34

p-adic field, 34
𝔓-adic field, 53
p-adic integers, 30–34
p-adic solutions, 14
pairing functions, 79
power character, 126
primality tests, 127
primitive recursive, 84
primitive root determination, 127

quadratic congruences with respect tc primes, 132
quadratic forms, binary, 134, 140
quadratic polynomials and forms, 39–46

recurrence:
  linear, 62
  order of, 63
recursion, functions obtained by, 84
recursively enumerable, 87
recursive set, 89
reduced form, 50
regular growth, 166
Rem (x, y), 79
residue class field, 59

sieve procedures, 146
singular dimension, 52
Strassman theorem, 54, 69

Tauberian theorem, 190

valuation:
  nonarchimedean, 34
  p-adic, 34
  𝔓-adic, 53

Waring's problem, 17
W-coefficients, 56
Weierstrass expansion, 55